VENUS

THE PATH OF BEAUTY

By Nick Kollerstrom

New Alchemy Press

Dedication:

to Alex,

who stood by me in time of trouble

Acknowledgements

Thanks to Mike O'Neill for originally producing the Venus-o-grams, to Jane Sutherland ('Chenka') for helping do the Venus-dance, to John Martineau for suggesting I write this, and believing in me, to Chris Sanders for doing the Kindle version, to Tunc Tezel in Turkey for permission to use his time-lapse Venus image (www.twanight.org); and to Stephen Windsor-Clive for substantial assistance and advice.

Contents

1

Her Perfect Motion

There is one planet which weaves out its dance around us in perfect beauty, harmony and proportion. Understanding this can even change your perception of the universe. It's a bit of a secret of our solar system, overlooked by astronomers. The latter are keen on their high-tech gear, and like looking at very distant objects, but alas tend not to notice the marvelous meaning of patterns woven by Earth's next-door neighbor.

In this book we will be exploring something a bit more permanent, which can reveal something both good and true about the kind of world we live in. Modern astronomy is largely fascinated with things which are invisible to the naked eye, objects which only ultra-powerful telescopes can discern.

The journey you are about to take, however, is more old-fashioned. For a start it is sensible astronomy, it concerns the way things look like from here, as perceived by our senses, and secondly it deals with what astronomy used to be about: ideas of beauty, proportion and harmony. It sings the music of Venus!

In these pages we will meet the Greek idea of *Kosmos*, a proportion in the total scale of things; an affirmation that the totality, the whole, has beauty. Before the fifth century BCE the Greek word Kosmos meant 'adornment' or 'pretty', as in lipstick being

Mandala of 8-year Venus pattern

'cosmetic' - it's actually the same word. We are going to try to re-access that meaning by looking at the perfect pattern that one planet weaves around us.

The word 'universe', or uni-verse means 'turning as one, but Venus draws attention to herself in a unique way. She, alone in the solar system, spins on her own axis in the opposite rotation to all the other planets. Everything else in the solar system moves and spins 'uni-verse,' turning-as-one, except for Venus! We need to converse a little on this subject (Latin verso, to turn).

The planets all travel the same way around the Sun in their orbits, and they all revolve, or spin on their own axes, in the same direction too. All except Venus, who spins the other way as she goes, thus setting up graceful patterns in space and time that are both musical and geometrical in their essence. Portia, in Shakespeare's *The Merchant of Venice* says: "The man that hath no music in himself, Nor is not mov'd with concord of sweet sounds, It fit for treasons, stratagems and spoils; The motions of his spirit are dull as night, And his affections dark as Erebus: Let no such man be trusted."

Modern astronomers need to mull over this stern judgment. When one can follow them, they tend to be dealing with chaos, instability and black holes, not to mention the inscrutable math equations. In contrast, we here endeavor to comprehend how beauty and proportion can be, just occasionally, part of the design of things. O yes, you will say, the surface of Venus resembles a tortured, hellish landscape, like some suitable venue for the damned. Well maybe, but for a start no-one can see it, we can only ever be shown a false-color reconstruction of its surface, and can we here begin with what is visible? Astronomers have been giving us a depressing cosmology suitable for a race of no-hopers and it seems to me that they are becoming mere slaves to their computer hardware.

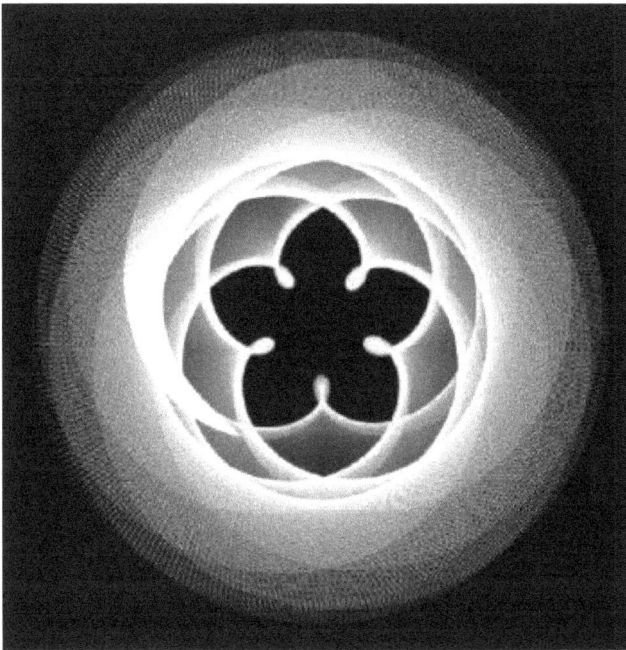

Figure: The path of Venus as it moves around Earth fixed at the centre, as if we are looking down from above. It's computer-generated, and represents a period of something like fifty years.

Science, *scientia* means knowledge and the knowledge we need to acquire concerns the divine harmonies and proportions which Venus manifests more perfectly than any other planet.

Nowadays, as the night skies fade from our view due to light-pollution, little more than the Moon, Plough, Venus and Jupiter remaining visible to most of us. Many of us are emotionally starved of the *meaning* of the night sky. Bright lights accompany us until bed-time, so the rods in our eyes hardly get used (these are the parts of the retina used for night vision - they will adapt, if you give them half an hour, to the silvery tones of a moonlight scene, our own hi-biotech night vision). We were never meant to live like this and we can't honestly hope to find satisfaction in it.

Let's start by trying to honor the planet Venus, as she appears in the night sky – swinging between her magnificent dualities of Hesperus the Evening Star and Lucifer the Morning Star, the one which dies down into the sunset at dusk in the East and the other (for the early-riser) arising boldly before dawn in the West.

In June 2004 and then in 2012 there were transits of Venus (when Venus passed over the face of the Sun). These were events that no person then living had ever seen. They fell on a specific date and specific zodiac degree, and, get this, both four years earlier and four years later, Venus had the *same* zodiac degree on the *same* date - more or less. In fact, in four years Venus conjuncts the Sun five times, alternating in front and behind. Er, how did that happen? Some music in 5:4 time might be helpful here, to put you in the mood, Dave Brubeck's 'Take Five,' or the 3rd movement of Tchaikovsky's 'Pathetique' Symphony. We examine this in Chapter Six.

The next figure shows an image, a pattern, made from the heliocentric motions of Venus and Earth, by joining them with straight lines every few days. The Sun is at the centre. It looks very similar to the earlier image at the beginning of this chapter - which had Earth at the centre. That's quite a puzzle isn't it?

Earth & Venus orbits (straight lines connecting) by Mark Pottenger, Sun at the centre.

How come this image looks quite similar to the earlier two in this chapter? The first diagrams depict *relative* motion, of Venus around the earth: that means, we'd get the same image for Earth moving with respect to Venus at the centre. It shows relative motion, of one sphere relative to another. But this one here instead joins up the positions of Earth and Venus, every few days, as they orbit round the Sun: a later chapter will look at, how to do this.

Venus tablets from almost four thousand years ago[1] honor its ten appearances and disappearances per eight years, as one of the first calendrical experiences of the human race. Can you see how the 5:4 rhythm echoes that pattern? We're going to be exploring in this text how this rhythm links the motion of Venus to the solar year. These tablets were found in Chaldea (Iraq) and are one of the earliest astronomical texts - before that there were just eclipse records and omen texts. They were inscribed a few centuries before the twelvefold division of the heavens appeared, in Greece and Babylon. Chapter 10 looks at this.

This text will introduce you to new harmonies that you hadn't heard of before. As women are often judged by their proportions, it's time to realise that the planet paradigmatically associated with love and beauty, our closest neighbor, our planetary partner, itself manifests the most exquisite proportions.

She's the Key. Astronomy concerns patterns woven in time and space, and a new starting-point emerges for its study, beginning

---

[1] 'Venus tablets from almost four thousand years ago:' .... See Michael Baigent, *From the Omens of Babylon'1994*, p.59: the Venus tablet of Ammisaduqa from 17th century BC.

not just with what is simple, but, more importantly, with what is perfect. But, what is, what can be, perfect? We here try to brush up this out-of-date concept, and blow the dust off it. The meaning of perfection will emerge from studying the Earth-Venus-Sun patterns! Patterns that are at once heliocentric and geocentric. We move between these two reference-systems, one centered on Sun and the other, here on Earth, the latter being what we experience.In what sense did this only recently become feasible? That will unfold, in due course. Let's just say that in the 1960s the rotation rates of Mercury and Venus - a vital part of the story – came to be discovered. So astronomers haven't had that long to assimilate their implications. Or, if you think that's loads of time, one could say that they aren't very interested. The comparatively tiny planets Mercury and Venus have enormously slow axial rotations, with the former having days longer than its year – while, in contrast, the vast, giant planets Jupiter and Saturn hurtle round unbelievably fast, having only ten hours per day. Have you ever wondered why this should be so? You probably think that astronomers have some 'explanation' of this. Well, maybe they don't! Let's try looking with fresh eyes – and, learn to wonder.

In more superstitious times it used to be considered unlucky to dance around 'widdershins,' anti-clockwise, the way the solar system actually revolves, *uni - verse*, turning-as-one. Find a friend and turn round with them, *con-verse*, turn together, perhaps as our Moon revolves, facing ever earthwards. Hold hands and spin around, or try both turning on the spot. Now do both at once! Thereby we can experience orbital motion and axial rotation.

We're now at the Dawn of the Space Age in this new millennium – but, alas, British expenditure on astronomy has sunk to only around two pints of beer per taxpayer a year. This little work is an attempt to show why Space does really matter to us, and how Harmony is generated in the realms around Earth. We're talking about our local neighborhood Space, not some distant galaxy far away.

The numbers that count

We all suffer from a *nausea mathematica* these days, of having too many numbers that we'd like to forget, a-jumbling in our heads. In the text that follows, there will only be numbers that have *meaning* and are significant in terms of the world-harmony. Other more factual numbers, the kind that one finds in astronomy books, what one might call 'data' (it means, that which has been given' in Latin, its in the past tense) are given at the end.

2

The Merry Waltz of Mercury

Let's start with the first planet, Mercury, and some first principles. Also the first numbers, 1,2 and 3. Its harmonies are easier to grasp than those of Venus. Venus' motion is like a grand symphony, where we can hope to experience one part or another at a time, but little Mercury just dances a merry 1-2-3 waltz all by itself. Let's assume that you are refusing point-blank to believe anything about harmony and proportion in the solar system. Let elusive Mercury will be our guide – after all, he was traditionally the magician / trickster. Have you ever seen him in the sky?

Imagine living on a planet where the stars all go round three times in the sky, every day. That's what you would see, on Mercury. You'd also see the Sun going exactly twice around the zodiac of stars, in the same day: *Mercury spins three times on its axis in two of its years, experiencing one day.*[2]

This is rather mind-wrenching, and give yourself a year or two to mull over it. Way back in the 1950s it was all so simple: Isaac Asimov wrote his classic 'I, Robot' about a bunch of robots that went to Mercury. The first planet was permanently locked into facing ever sunward, the scientists all agreed. The robots in the story noted how Mercury's 'terminator,' a line on the ground with

---

[2] Mercury spins in space once per 58.65 days, orbits the Sun once per 87.97 days, and has a 'day' of 176 earth-days: these are in a 1,2,3 ratio.

Day on one side and Night on the other, was forever fixed, immobile, upon the rocky surface!

This belief came about because, whenever astronomers were able to look at Mercury's surface thru their telescopes - which they could only do in one part of its orbit when it was furthest from the Sun - they always seemed to see the same side facing them. Ha! Mercury had them well tricked.

We now need to focus on the three concepts of day, year and axial rotation period. (i.e., the time it takes to spin around once in space). This is a kind of dry run for when we come to Venus, where things get more complicated. Don't worry, it won't hurt a bit.

For us on earth, the axial rotation period, or the time it takes for a star to rise again, is called the sidereal day. It more or less coincides with our day, only less a few minutes. So the Earth-star day and the Earth-Sun day are much the same, within a few minutes. Only astronomers bother about this difference.

On Mercury there is a beautiful threefold harmony, a 3:1 differential, between its rotation against the stars and its own day. But, there are more startling tricks which Mercurius has up its sleeve...

On Mercury, Asimov's robots would have witnessed a display stranger and more wondrous than any sci-fi writer dreamed up. Standing on the surface of Mercury, they would have seen the Sun rise and climb up into the meridian, then stop and turn back - and set, then rise again, before then tracking right across the sky. Mercury's rock-and-roll sunrise! It is subject to ferocious accelerations during its highly elliptical orbit, and these somehow put that strange warp into its day. The year of Mercury is of course its orbit-period around the Sun – if it had seasons we might want to define its year in terms of these, but it doesn't really.

The same part of Mercury always faces Sunward every alternate perihelion. (when it is nearest to the sun, from peri, near and helios, sun). It is thereby manifesting an incredibly strange relation to the Sun. It was nothing like what scientists had surmised from their materialistic theories, but more like a dance. If that wasn't enough, it also subtly manages to point Earthwards at key moments, so that when it is briefly visible to Earth it is usually manifesting the same side or face to our telescopes. You can see here that the celestial dance of tiny Mercury is one of *interconnection* and *linkage*, as it relates to both Sun and Earth. This is the bit that astronomers tend not to like: they don't like it because it *means* something.

Mercury's path around Earth, by Keith Critchlow

So, why don't astronomy books ever mention the charming 1 - 2 - 3 dance with which tiny Mercury goes a-whirling round the Sun? It's simple, beautiful and harmonic. It isn't taught *because* it's simple, beautiful and harmonic. Let that sink in. Ask astronomers about Einstein and they are liable to start warbling on at length about some tiny shift in a fraction of a degree per century in the plane of Mercury's orbit. Well, no thanks! We're trying to stick to more sensible things, a bit more visible if you know what I mean.

Let's turn now to an Earth-centred diagram by the geometer and architect Keith Critchlow, showing the three 'loops' Mercury makes in the course of one year (The Sun's passage is also shown, during that year). It weaves a triangular pattern. One should not start something, say astrologers, when mercury is 'retrograde.' 'Retrograde' here means, that it appears to move backwards against the stars, for a while. For the outer planets we can observe these periods of backwards-motion, but for Mercury and Venus, we never see these loops in the sky.

That diagram is a view from above. It shows the retrograde loops which occur during its 'inferior conjunctions' with the Sun. We'll meet inferior conjunctions later on. These important events happen when Mercury or Venus are situated between us and the Sun, which is when they come closest to Earth.

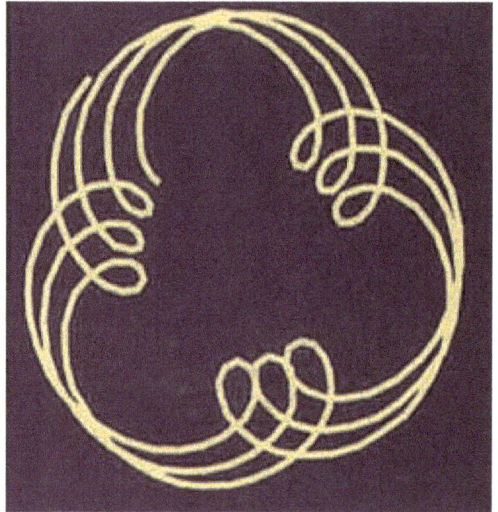

Figure: Mercury's dance around the Earth, in three Earth-years.

We turn next to the orbits, i.e. motions around the Sun, of Earth and Mercury, and picture them as circular: What astronomers call the 'mean' orbit is a kind of average, even though Mercury zooms about at hugely varying solar distances. Over time a planet's wobbly orbit would slowly build up a thick spherical 'shell' – the mean orbit is the simplified version of all that, reduced to just one circle. John Martineau's discoveries were done using this 'mean' orbit concept, and we come across some more of these in a later chapter.

He began in the early 1990s finding these, starting off with Mercury, while taking an M.A. at Keith Critchlow's School of Architecture. It was seeing an astronomy book by Joachim Schultz that got him started: that book took a *geocentric perspective* on the heavenly motions, and as we'll see that is what counts.

He found out that the mean orbits of Mercury and Venus are defined by *pushing three coins together*. The diagram here fits to a staggering 99.9%. It somehow reminds us of the 1-2-3 rhythm of Mercury's motion that we started off with - but here it appears in space, not time.

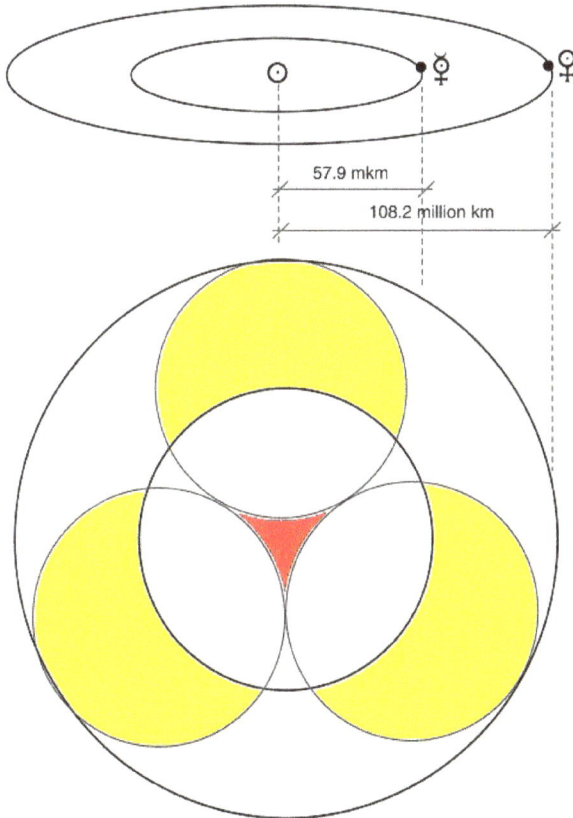

Venus and Mercury orbits, with three circles: John Martineau

In a later chapter we look at some more of his diagrams, but for now we move to an image that looks completely different, a diagram by the Steiner astronomer Joachim Schultz. Mercury here looks to be buzzing around like a demented bee. I mean, where is it going? One feels rather alarmed that it might have lost its bearings. This is a one-year period of its motion, as seen from the Earth looking towards the Sun. He did this for the year 1961. He died in a motor-bike accident and his classic opus had to be finished off posthumously by another. It has remained in print ever since, which you cannot say for many astronomy books.

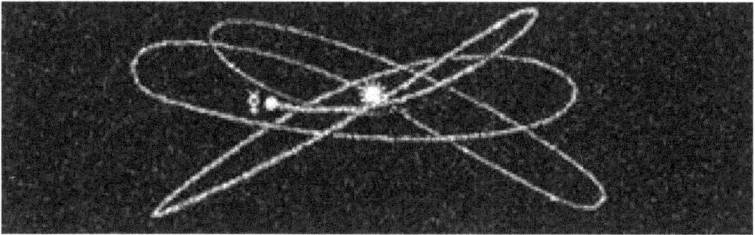

Mercury's path, with the Sun stationary, for the year 1961: J. Schultz 1986 p.144, after JM 2001.

If we call this a 'Schultz diagram,' it expresses the difference between Mercury and solar co- ordinates.: subtract the two and presto! out it comes.[3] Here is another one for the year 2006. I've added in the days when someone reckoned they could see Mercury, in the morning and then in the evening. For a couple of weeks in February and then in June it was visible in the early evening (left hand side) then later on in parts of August and November it was seen in the morning (shown here on the right hand side). We see a line going right across the Sun this year, because there was a 'Mercury transit' when astronomers were able to see it moving

---

[3] Here, Joachim Schultz has subtracted the astronomical coordinates of right ascension and declination for [Sun – Mercury], which puts the Sun immobile at the centre.

across the Sun. Astronomy clubs need to make these diagrams and note on them the days when they espy nimble Mercury.

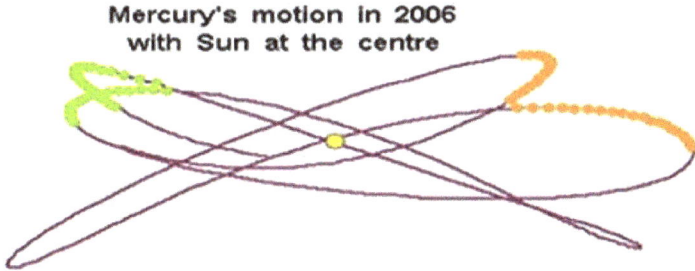

A similar diagram, showing the periods when Mercury was visible.

By now you may be feeling confused, because of the several different points of view that we have taken. We started off from a robot on Mercury watching its sunrise, then we computed some key ratios, then looked at the loops of its retrograde motion, and now have been shown what the view of its path from the Earth would be like, if we could see it moving about close to the Sun (which we can't). Cosmology does involve us in taking different viewpoints, and there is no one correct view. What matters here, what is important, is the perfect harmony.

'Resonance' the astronomers say. Resonance happens when you put a tuning- fork onto a grand piano, and then suddenly hear the note grow louder. It's a transfer of energy between two systems oscillating at the same frequency; or they can be oscillating at simple integer ratios, for example an octave, it will still be resonance.

How is this relevant to astronomy? Astronomers are generally agreed that the 'lock' between Mercury's day and its solar orbit in a 2:3 ratio is somehow produced by resonance. They have to agree because there's no other solution in sight.

For ages astronomers said the same thing about the Moon, which keeps its same face forever facing the Earth so that no Earthbound telescope has ever seen 'the dark side of the Moon.' This was caused by 'resonance,' they explained, adding 'spin-orbit coupling,' and what could one say to that? There was a meant to be a bulge on the Moon and Earth's gravity thereby holds it facing Earthwards. This is important because both Venus and Mercury 'face' earthwards under different conditions so let's try and get a focus on this.

Luna revolves by herself every twenty-seven days, so her same side always faces us. The Moon's orbit-period and axial rotation are synchronized, they are locked in together. Is this due to Earth's gravity's pull upon her? When astronauts went out to visit the Moon, back in the 1960s (Yes they really went!), they discovered that it had a large bulge, a huge one in fact, but it was on the wrong side - on the far side, away from Earth. They found there was something extra-dense beneath the huge frozen lava seas of Luna that existed only on the Earthward side, no-one knew why: so the whole post-Apollo argument as to why it faced earthward came to hinge upon some rather peculiar dense subterranean masses, an arcane secret of Luna's past that no-one was ever likely to de-crypt.

Luna is so far away that she isn't as such held by the Earth, like a satellite should be, but it sojourns round with us, pulled always more than twice as strongly by the Sun than by Earth. An astronomer would say that its path is always concave with respect to the Sun. Because of this, anyone wishing to explain why Luna faces earthwards has to start off by assuming that she was once much closer to us, when the gravity pull was stronger. We needn't go further down this path.

So we have an explanation based upon resonance and gravity-theory, which should have applied more forcefully to Mercury. If, long ago the latter was once semi-molten (so the argument goes), then tides from the Sun's mighty pull ought soon have damped out any independent rotation it might have had, to make it end up facing ever Sunwards.

How did it go so wrong?

Sources: *Rhythmen der Sterne* Joachim Schultz, 1963. Keith Critchlow, *Time Stands Still'*, 1979, p. 160. John Martineau, *A Book of Coincidence* 1995.

3

The Dance of Venus

A Double-pentagram, woven by solar conjunctions over eight years

Every eight years, the double-pentagram Venus is woven in the sky around us. The pattern is formed by its meetings with the Sun. This period is the basic quantum-unit of Venus' motion. Five times does it come nearest to us, as shown by the small pentagram: its 'inferior conjunctions.'

These are the *perigee* and *apogee* positions of Venus. 'Peri' means 'close to' in Latin and 'gee' is from *geo*, the Earth. The two pentagrams give the nearest and furthest positions of Venus from the Earth - just as the Moon reaches its apogee and perigee positions each month.

As it weaves out these two pentagrams every eight years, they stay in synch with each other: whenever Venus meets the Sun when it is closest to the Earth, four years later to the day it will again meet the Sun when furthest away, at the same point in the zodiac. The zodiac is punctuated by this eight-year fivefold pattern. Earth is at the centre - where she should be.

Thus ten of these celestial meetings chime every eight years. These nice old words 'superior' and 'inferior' allude to its going behind the sun when its far away from us (the idea that 'superior' was 'above' the Sun) then coming close and growing brighter as it swings in nearer to become 'inferior.'

Slowly these pentagrams in the sky revolve round against the stars, once per twelve centuries – exactly! So the number twelve turns up here. One pentagram is roughly six times larger than the other. As we saw with the Mercury-rhythms, astronomy books do not describe this marvelous double-pentagram. It was not 'known to the ancients:' thus *new principles of harmony* are being discovered in our lifetime.

In terms of a child's birthday, each four years it gets a 'Venus-return' as one of its 'many happy returns,' i.e. it comes back to the same position in the sky. By twenty-four years of age this has moved a little out of synch, the 'return' being six degrees away.

Here the table shows how these two pentagrams pan out in time, so you can see how closely the pattern repeats, every four years – almost to the same degree and same day. We can here follow the four-year repeats of the pentagram pattern in the sky, and see how close they are.

<u>Sun-Venus conjunctions 2010 – 2020</u>

Capricorn	Scorpio	Leo	Gemini	Aries
11.1.10	28.10.10	16.8.11	6.6.12	28.3.13
22°	5°	23°	16°	8°
11.1.14	25.10.14	15.8.15	6.6.16	25.3.17
21°	2°	23°	16°	5°
9.1.18	26.10.18	14.8.19	3.6.20	26.3.21
19°	3°	21°	14°	6°

Our Moon manages an even higher precision: every nineteen years at your birthday the Sun and Moon arrive in the same position in the zodiac and in the sky as when you were born, and that will keep going right through your life (on your 38th and 57th birthdays). That's called the Metonic cycle). Maybe Sol and Luna are more concerned with the measuring of Time.

The Heart-and-Rose Mandala

Venus' Earth-centred dance over about 35 years

The famous Earth-centred Venus heart-and-rose mandala shown above first came into use in the early 1980s, it emerged from home computer programs. It seems to have been of little or no interest before computers started to draw it. I knew the California astrology publisher Neil Michelson who published all sorts of planetary mandalas in the 1970s, but this one of Venus was not among them. These diagrams work well because Earth and Venus have nearly circular orbits. Earth comes just three percent closer to the sun in midwinter than it does in midsummer, and Venus' distance changes even less – it has the most perfectly circular orbit of any in the solar system.

In contrast, the inmost planet, Mercury, has a very elliptical orbit - like Pluto, the outermost planet. Earth and Venus rejoice together in their nearly- perfect circularity, unusual in the solar system. That's why we can ignore the elliptical motions here.

If you just saw the Moon's path, from above, it would look like a circle. We only notice it being elliptical because Earth's position isn't quite in the centre. Let's call that a fifteen percent eccentricity Venus' Earth-centred dance over about 35 years because the Moon comes fifteen percent closer to us at perigee. Earth has a mere three percent eccentricity going around the Sun – that's how much closer we are at perihelion than at aphelion. Venus is even less, more like one percent. Neptune's orbit is the only other having a comparable circularity to that of Venus. How can that be, how does the Universe come up with such balance? Nobody really knows.

Isaac Newton saw the nearly-perfect circularity of planetary orbits as evidence of the 'divine arm' that set the whole show a-rolling at the start; otherwise, he said, they would all have highly eccentric orbits like the comets. That is a tempting argument, which has grown a lot stronger since 1997 when astronomers have started checking out other solar systems on distant stars. What have they found so far? Mostly they have found large planets with highly eccentric orbits.

If this carries on, we may have to wonder, whether the neatly-set out arrangement of our solar system, almost circular and nearly in the same plane, was the result of a creative process that involved harmony rather than chaos, proportion rather than chance. We may not want this, after all who wants Big Daddy back again with his Divine Arm, for goodness' sake? Plato viewed the planets as divine on account of the mathematical beauty of their circular motions, and saw circles as being the most perfect shape.

There is an old, well-forgotten paper by two British astronomers who checked out 'resonances' in the solar system, and concluded that there were more there than should exist by chance.[4] And what is resonance? Well, it happens between Jupiter and the asteroid belt: huge Jupiter causes the asteroids to concentrate in certain orbit-periods that are simple fractions of its own. Conversely, there are some gaps in asteroid belt orbits that have been evacuated by Jupiter. Then, as we saw earlier, there is a possible resonance argument concerning Mercury and the Sun. We'll return later to the question of whether it can likewise be applied to Venus. We're not quite there yet, have patience.

A small group can have fun weaving out the dance of Venus. As one whirls around performing this dance, the diagram shows how to speed up and slow down. The spacing of dots indicates the speed, over one single 'heart.' Five of these comprise the total 'rose' that is woven. While coming in close to the centre one slows right down, then speeds up around the outside. The second diagram shows how far Venus goes in (arbitrary) equal intervals, so it

---

[4] Archie Roy and M. Ovenden, *Commensurable … Motions in the Solar System*, Monthly Notices of RAS 1954, pp.114,232.

indicates just how much dancers slow down and speed up as they move around. It's difficult – in fact, it's very hard, few people manage it!

When farthest outside at each loop one passes by the 'superior conjunction,' and these diagrams do not show where this is located. As we saw with Mercury, Venus draws closest to Earth while it is retrograde and at its inferior conjunction. It moves retrograde less than any other planet, a mere seven percent of the time, so only one in 14 people are born with Venus retrograde.

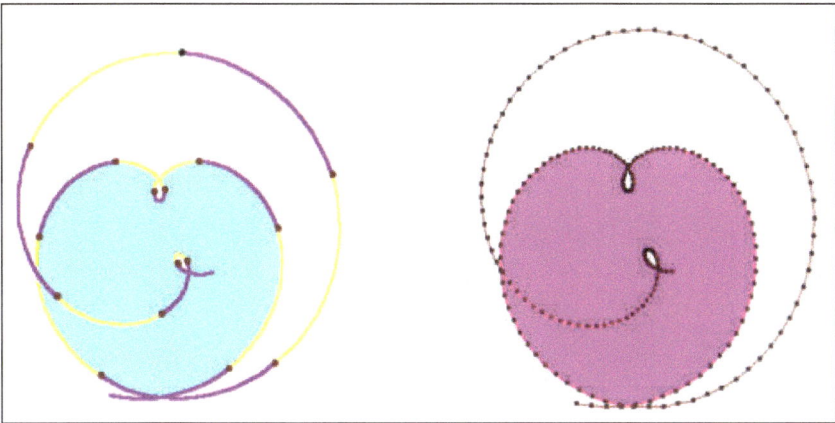

To draw a heart: two years of Venus' motion, showing its varying speed around the Earth

Earlier we saw how these motions created two sky-pentagrams, but now we have a different picture to show the geocentric motion. Again it is an image not known to the ancients, it's a new experience. People like to experience it, because it's a mandala, which means that it promotes our wholeness, our integral experience, which we all need. We don't just want abstract diagrams, of orbits circling the Sun. This picture is in relation to *us,* pointing towards the Earth. It shows something special about the pattern of interlinkage between Venus and Earth. Dancing the path of Venus helps us understand how our closest planetary neighbor actually moves in space around us.

Her Vital Statistics

Synodic period	583.92 days
Axial rotation	243.02 days
Year (sidereal)	224.70 days
Day	116.8 days

Corn Marigold: 13 petals

4

The Divine Proportion

Golden Ratios

Perfection isn't an easy concept, I admit. One can get habituated to chaos and violence. A good place to start here is the so-called Golden Ratio So let's turn off the TV and try to meditate calmly upon the regular pentagram figure, because of all shapes it does best express the golden ratio. This ratio is known as phi, represented as Φ. All of the pentagram shapes make phi-ratios! How is that?

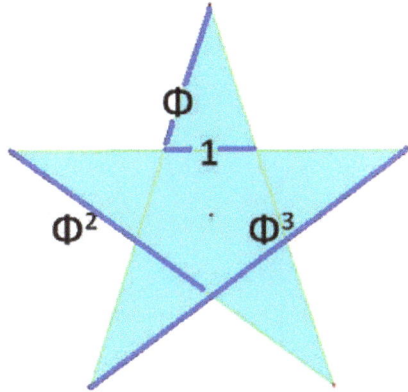

A regular pentagram, showing how it contains powers of the Golden Ratio

The Golden Ratio divides a line so that the whole and parts are related to each other in the same way. There is a large part and a small part:

$$\text{Large Part} / \text{Small Part} = \text{Whole} / \text{Large Part} = \Phi$$

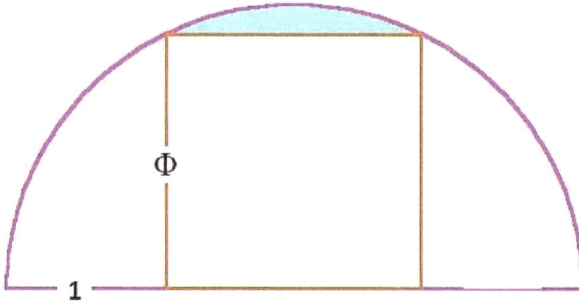

Figure: A square in a semicircle makes the golden Ratio

The idea of perfection in the golden ratio comes from the special way in which it reflects back on itself. A square added or removed from a golden rectangle simply creates another golden rectangle. We can construct phi by fitting a square into a semi-circle (above).

A studio window built on this design might help you see things in divine proportion. Pi comes from 'squaring the circle,' but phi comes from a square in half a circle!

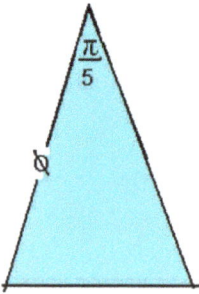

Mathematicians allude to the 'divine triangle' and this uses the pentagram angle of one-tenth of a circle. We can express it using both pi and phi, as shown: the sides make phi the golden ratio, while the angle at the top is $\pi/5$ (radians, i.e. 36°). One should feel a tingle of awe when those two numbers, π and Φ, come together. Also, here's a nice

$$\Phi = 1 + \cfrac{1}{1 + \cfrac{1}{1 + \cfrac{1}{1 + \cdots}}}$$

equation, which seems to give a clue as to how the Golden Ratio is really zooming towards infinity:

Bottichelli's immortal masterpiece, *The Birth of Venus* is the world's most popular painting. We devote a whole chapter to it! But for now, wek merely notice that in it, the entire attention of the viewer is focused on to the navel of the goddess, and that coincides with level horizon, which divides the whole painting in golden proportion vertically. It's so exact he must have done it deliberately.

If you measure your height, then measure how high your navel is from the floor, and divide one of these by the other, you'll come out somewhere near the Golden Ratio.

Nowadays girls expose their midriff quite a bit, so our attention is so to speak focussed on this ratio. Your navel button divides your height in the golden ratio.

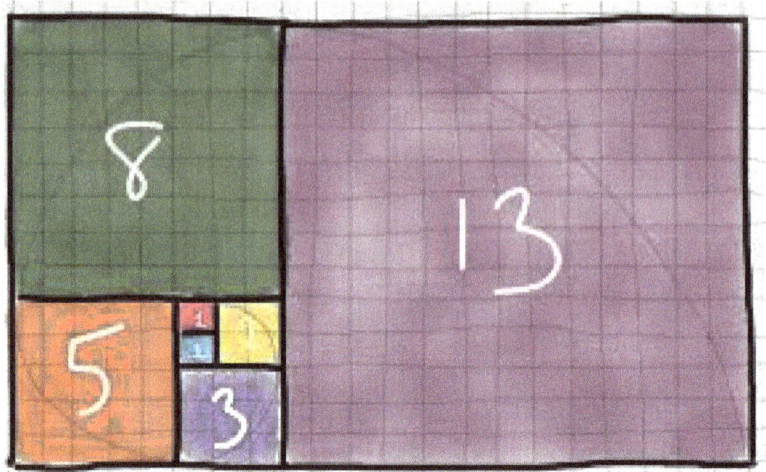

A nest of 'Golden rectangles' can be drawn, which spiral inwards as shown. They never reach the centre, and its shape remains unchanged however much it is magnified!. A curve just touching all these rectangles is called a 'phi-spiral.' And this brings us onto the Fibonacci series, a series of numbers whose adjacent terms form ratios that move ever nearer to the Golden ratio. Like the tortoise in Zeno's Paradox, they don't ever quite get there. The sequence is 1, 1, 2, 3, 5, 8, 13 et ceteera, with each term the sum of the two previous ones. Five plus three equals eight, and 8/5 is closer to Φ than 5/3. As the numbers increase, the ratio between them moves nearer towards the pure, golden ratio.

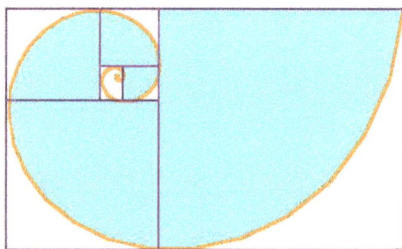

A Fibonacci-buff can pick up a pineapple and show you its 5:8:13 ratios, woven between its spiral whorls going five in one direction, eight in the other and thirteen vertically. While munching its clear, sweet taste we digests the implications. Pine cones generally shows a similar pattern, either 5:8 or 8:13.

Sunflowers use higher Fibonacci ratios, in the number of spirals coming out from the center of the flower, first turning one way then going the other, but these aren't essential to our story. Most plants on Earth use the numbers 5, 8 and 13, and we are here discovering how these are Venusian numbers.

Taking the sequence 1, Φ, Φ^2, Φ^3, etc. where Φ phi is the Golden Ratio, each one is the sum of the two previous ones, just like the Fibonacci series. We go up this series by multiplying by Φ at each step. So, addition and multiplication strangely come together in the golden section series. Some believe that the facade of the Parthenon in Athens was built in this ratio (I'm never convinced). Before the Turks blew it up it must have looked awesome as a temple to Athena. The name 'phi' was chosen for the golden section in the 20th century in honor of the ancient Greek sculptor Phidias, who was involved with decorating the Parthenon.

A Fibonacci whirl

Pentagrams are not found in rocks or crystals but they burgeon in living things and especially sea-creatures. We saw how there are several ways in which phi can be found, in a pentagram, it's the only figure made entirely of golden ratios - all its sides are 'golden' to the others. Christians like their four-sided Cross and Jews like their six-sided Star of David, but women wearing five-fold pentagrams, they must be trouble!

It's sides make the ratios 0.618, 1.618 and 2.618, that is Φ^{-1}, Φ^1 and Φ^2. Its corner angles are a one-tenth division of the circle, which is the angle used in the DNA molecule: its spiral turns once round per ten steps, so the 'molecule of life' is continually expressing the golden proportion.

When you have got a sense of how the different pentagram sides reverberate to phi the divine proportion, then we can move on. This is a proportion found in the plant realm, but also in the dance of Venus. To see this, we need to take an earth-centred or 'geocentric' viewpoint. Only the viewpoint from Earth will give us these 'vital statistics' of Venus.

Years in synch

Let's compare the two years, of Earth and Venus. Their ratio spells phi to within one percent. Taking the Venus synodic period (how long it takes to be behind the Sun again) of 19 months and comparing this to the Earth year, it gives phi to half a percent.

Venus' synodic cycle: Earth's year = 8/5, to within 0.08%

Earth's year: Venus' year = 13/8, to within 0.03%

So, the planet of love, beauty and harmony once again expresses its connection with Earth; just as it did earlier when we saw how the same part of its surface always pointed earthwards at each nearest approach. Its celestial music is deeply oriented towards earth, that is to say it can only be heard from a geocentric viewpoint, from Earth.

We were reminded of the Golden Ratio by the starry pentagrams which Venus inscribes around Earth each eight years, and now it has turned up as a pattern in time. This is mostly new research. Astrosophist Hazel Straker first described the double pentagrams, and archaeo-astronomer Robin Heath first described golden ratios in the orbit-periods. It seems that the divine harmonies of Venus only started to be apprehended by humanity in the late twentieth century, which is why they aren't in the astronomy books. Perhaps the 'prophetic soul of the wide world dreaming of things to come' wants us to start experiencing harmony and beauty once more.

Fans of Buffy will apprehend that merely seeing a pentagram indicates that some otherworldly being, probably some unsavory vampire-type, is likely to appear. Even worse, it has been the emblem of the US army wherever it wreaks its mayhem. Mephistopheles in the story of Faust declared that a pentagram on the wall 'causes me pain,' so what the devil was going on? For these reasons let's stress that the double-pentagram is the true emblem of Venus. There is something creative about the pentagram, as if something unexpected may appear. I like the phrase 'the irregular and vital beauty of the pentagram' – from the New York architect Claude Bragnon.

The patterns in time we looked at earlier are closely equal to these Fibonnacci ratios: That's how the years of Venus and Earth form lovely steps in that sequence – which is based on the golden Ratio - exact within ten-thousandths.

The piano keyboard has the thirteen-note chromatic musical octave consisting of eight white keys (whole tones) and five black keys (sharps and flats). The corn marigold and chamomile are flowers having thirteen petals, delphiniums have either thirteen or eight petals, and buttercups and edible fruits all have five petals.

A briar rose has five petals, but we are hardly able to see that fivefold pattern in modern bred roses; except maybe, just as the bud is opening. As the leaves of a rose move up the stem, each one turns round three-eighths compared with the one before. That's part of the Fibonacci series, or rather 5 to 8 is, counting the other way. For comparison, that turning-ratio is 1/3 for beech trees, 2/5 for oak and apple trees. 3/8 for sunflowers and 5/13 for willow trees. It's called *phyllotaxis*.

These number-pairs continue within the flowers, moving up to higher ratios of the Fibonacci series: the daisy often has 34 or 55 petals, which is often the number of spirals in a sunflower, or it can be 55 and 89, where one spiral goes clockwise and the other anticlockwise. These higher ratios move ever nearer towards the Golden Ratio. These are *flower-numbers*. The commonest ones, 5, 8 and 13 have a marvellous Venusian meaning.

Her Path of beauty

Venus, two synodic cycl
one line per week

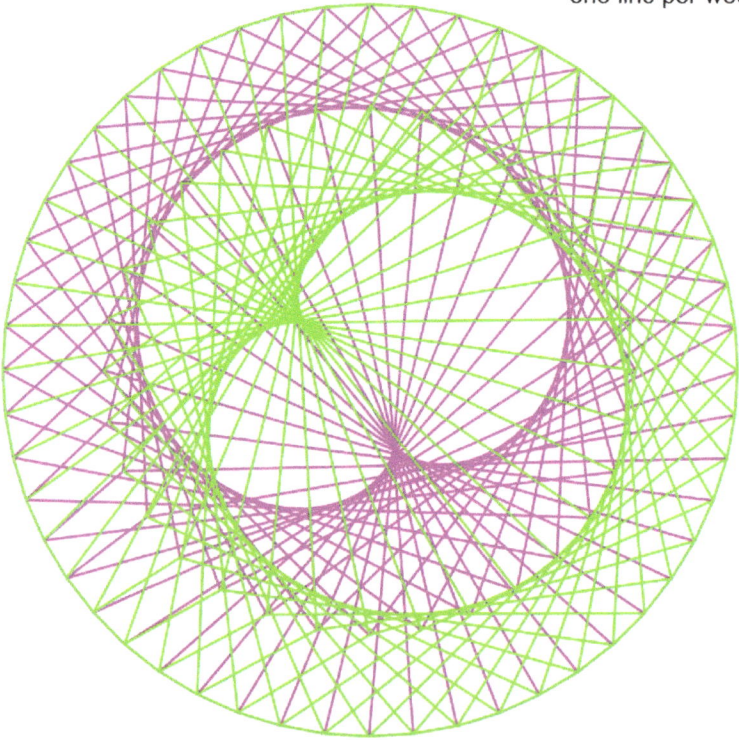

This image depicts two synodic cycles, showing lines joined up between Earth and Venus, with Sun at the centre. A complete diagram goes through five of these cycles to make a rose-pattern, as we saw in Chapter One. But, just two or three may better show the graceful interaction of Earth and its nearest neighbor, as it weaves the fivefold pattern.

Draw two concentric circles, in scale to represent the Earth and Venus orbits (see Chapter Eight). Their years are exactly as 13 : 8, the Fibonnacci numbers - as we've just seen. Multiply those up by four, and that gives 52:32: we divide the two circles in that manner, so the outer circle gets 52 divisions, i.e. one per week.

Rule the shortest line you can joining marks on the two circles, and that represents the inferior conjunction, when the two planets are closest together. Then, keep going, ruling the next line and the next, and when you have gone round the circle one and three-fifth times, that is one synodic cycle. So, the pentagram pattern comes in here. Eventually this will give us the full pattern, as shown on page 5.

With its sweet fragrance above and thorns underneath, poets have always associated the rose with Venus.

Wild Rose, by Karen the 'Graphics Fairy,' with kind permission

Source: Matila Ghyka, *The Geometry of Art and Life* 1977, Dover NY on golden ratio and human form.

5

The Music of Venus

"O body swayed to music, O brightening glance,
How can we know the dancer from the dance?"

W.B. Yeats

We've looked at the 5:8 and 8:13 ratios and now we come to Venus' last and most subtle rhythm of 12:13. That's the final one she uses for her complete symphony. But I have to warn you that it is a bit tricky. It happens within the eight-year period, as she weaves out the heart-and-rose mandala. To do this we have to focus on her axial rotation period.

We've seen how the geometry of Venus indicates a relationship with the Earth. Well, she here proves her affection again through the synodic period.

Synodos is a Greek word meaning meeting, and the synodic period of Venus, which is between eighteen and nineteen months, tells us how often she will grow into the bright Evening star, and then pass behind the Sun. It's an experiential interval between us and her and it's one-fifth of eight years. Earth and Venus come closest, so that we see her shining most brightly, each synodic cycle.

The period covers two Sun-Venus conjunctions, superior and inferior. As with the lunar cycle, Venus grows into her period of glory and then fades away, sinking further each day into the dusky turquoise of the sunset. Some say that this disappearance pertains to old Babylonian legends of the goddess Ishtar descending into the underworld.

Now, get this. Venus revolves just twelve times in space per eight Earth-years. That is her rotation relative to the stars.[5] When you are all by yourself, just spin around. That's your personal rotation in space. If you move round with somebody else then that is more like an orbit, you are both moving round a centre. Venus' own rotation period of 243 days is two-thirds of an Earth-year, slower than any other planet – and, she is spinning in the opposite direction. How in heaven's name did she manage that? And, she herself experiences thirteen years during this period. So in the five synodic periods (five hearts woven out) she goes eight times round the Earth, revolves twelve times in space and goes thirteen times round the Sun. That is the symphonic whole. [6]

Halleluja! I praise whatever Creator-God set this all up … rightly rotating to remembered rhythm.

Venus experiences 5.001 of her days per synodic period. A 'day' means how often a Venusian would see the Sun rise. We saw how in the last chapter, performing the Venus-dance, one comes to face the center four times per 'heart,' i.e per synodic period. And the center of that dance is Earth. This means that there is a lovely 4:5 rhythm that a Venusian would experience, in the interval of sunrises and earth-rises. From Venus one would see four Earth-rises and five sunrises per synodic period - rising in the West. Phew! We'd better take a break after all that.

Resonance? A step too far

There is a well-forgotten report from a meteorology department, mouldering away in some dusty old volume, by an Australian Mr

---

[5] 5 x synodic period => 12 x axial rotation period (to 99.9%).
[6] Synodic period => 583.92 days; its axial rotation takes 243.02 days, its year (sidereal) is 224.70 days, and its day is 116.8 Earth-days.

Bigg[7]. He discerned that each time Venus draws nearest to us, Earth's magnetic field becomes subdued. Venus is then nearer to us than any other planet ever comes. No-one is sure how Earth's magnetic field is actually produced. It varies in a lively manner from day to day and reverses at intervals through geological epochs – but, he found that it somehow, quietened down as Venus drew near. Thus the inferior conjunctions have a *geophysical* effect. This report appeared shortly before astronomers managed in 1967 to penetrate the dense clouds covering Venus's surface from our prying eyes, to detect for the first time ever her own rotation in space. A few years before that, they had detected that of Mercury.

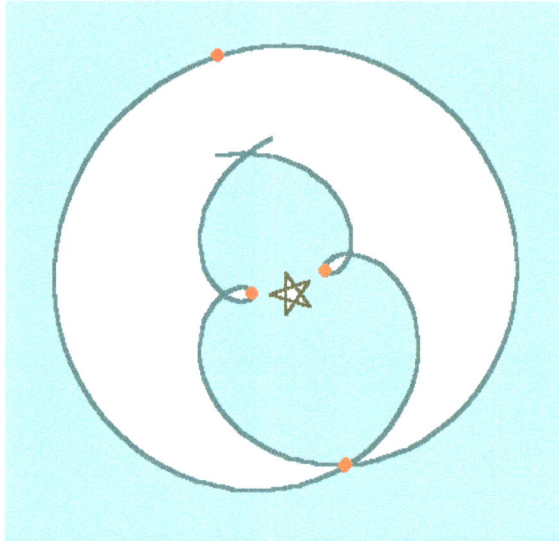

Figure: Two years of the Venus-dance with Earth at the centre, where the pentagram shows the 'stationary points' of solar conjunctions. The four dots show where the same side of Venus faces Earthwards.

---

[7] E.K.Bigg, 'Lunar and Planetary influences on geomagnetic disturbance' J Geophyss Res. 1963, 68, p.4099.

The same side of Venus's surface faces earthwards both at superior and at inferior conjunctions - that's ten times per eight years. So, cast your mind back to the Venus-dance. We had to face earthwards four times per heart, which works out at twenty times in all as we wheel eight times around against the stars. Try to feel it, feel the music of Venus. Here are two pictures of her dance, showing the positions where her same side faces earthwards: can you see how one fits into the other?

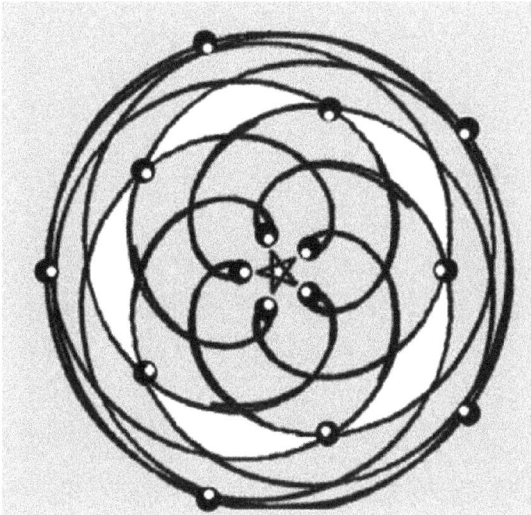

The same over an 8-year period, showing points in the orbit where the same side of Venus points earthward. (NB I only found 15 here, but there are twenty altogether)

'Resonance' the astronomers said. But resonance has to mean that a physical transfer of energy takes place. For example we've all heard of how tidal forces are created by the Moon and how this slows down Earth's rotation and pushes the Moon further away (don't worry, its only by centimeters per year). So a transfer of energy and momentum is taking place and so we can just about believe that this could account for Luna's lovely face facing ever earthwards.

Or, suppose I were to hold forth about how the back and front of the Moon are completely different, with the side facing us being a surreal dreamscape with huge, dark once-molten lava seas, with its long silvery rays a-shimmering and its strange sinuous rilles meandering about as if they were once river valleys (which they never were); whereas the back of the Moon is quite uninteresting with chains of mountains and no lava seas, and merely looks like so to speak the back of someone's head; and that poetry might hardly have existed if the front of Luna resembled its behind, and.... Well, I doubt if you'd be convinced. A moderately convincing physical argument does exist as to why it is locked into facing ever earthwards, so we don't really need poetic or theological arguments as to why it does this: resonance will do.

Then we saw how with Mercury astronomers are happy to have resonance account for the 2:3 music between it and the Sun. Neither you nor I (I guess) can handle the equations of its elliptical orbit, so we take their word that such interaction does produce such a result.

A Passing Biff ?

Astronomers sometimes say Venus received a 'biff' by some passing comet or whatever, which knocked it into its reverse-rotation mode. Can you believe it? The modern theory about how the solar system was formed by a 'pestilent congregation of vapors' as Hamlet mused to himself (with man as the mere 'quintessence of dust') has everything revolve and rotate the same way as the primeval solar system gradually coalesced and cooled down. There are major problems here. For a start Venus' axis of rotation is fairly vertical: it doesn't look as if it's been knocked about. Secondly, Venus has a more perfectly circular orbit than any other known planet, hardly what you'd expect after a big knock. Thirdly, does its orbit diverge from the plane of the ecliptic as one might tend to expect from a bump? No! And fourthly, were its rotation due to impact, would one expect it to be revolving more slowly than anything else in the solar system? One can hardly have the exact musical- harmonic ratios of 12:13 and 5:4 resulting from collision with a comet!

Not only will a cometary impact not work, but resonance won't work either. Venus is fairly spherical, it has no great bulge anywhere, so Earth's pull can't have gotten a grip on it to alter its rotation, and there is no tidal means whereby a lock could have been established onto its rotation, to slow it down or speed it up. So it seems that a dance is going on that physics can't currently account for. The astronomer Kepler wrote about Harmonices Mundi the 'Harmonies of the World' and, unfashionable thought this may be today, Venus is showing us just this.

Venus' axial rotation demonstrates *quantum relationships*. Let's go through them: first there is the 2:3 ratio of its axial rotation to Earth's year (99.8%); then there are the five Venus-days per synodic cycle (99.98%) plus the four 'Earth-days' per synodic cycle (99.96%) - i.e. on Venus one would see four Earth-rises in this period, and finally there is the 13:12 ratio between its axial rotation and year (99.8%)[8]. It has required the most exact calibration of the Venus-rotation as performed by space-equipment in recent decades, to obtain these awesomely precise, interlocking synchronies.

Every eight years the Venus-pentagrams shift around by some two degrees. The two pentagrams of Venus revolve majestically against the stars, exactly once per twelve centuries (to 99.9%, or 1199 years). We have earlier come across the number twelve, in connection with Venus' rotation against the stars. The next chapter looks at the effect of this twelve-century period, in which the Venus-pentagram revolves.

---

[8] J.S.Lewis, 'Venus and Earth: another dynamical connection?' Astronomy & Geophysics Aug 1998, p.4.8

Nine times slower than the Moon

These numbers can sound so bewildering. Put your feet up on the verandah, listen to the rustling of the trees, sip some elderflower cordial, and mull over them. That special number, the lynchpin of Venus' symphony, which gives her such majestic slowness of spin, is the *fifth power* of three:

$$3 \times 3 \times 3 \times 3 \times 3 = 243$$

It revolves in space more slowly than any other planet. Mars spins on its own axis once a day, much like Earth, and Jupiter twice, so how did Venus get to have this special rotation period, six hundred times slower than Jupiter? We've already suggested an answer in terms of the elegance of the dance that the ballet dancer performs, to give perfect poise and grace to her motion (Chapter 3), but you may not have been quite happy with that angle.

Comparing Venus with the Moon, the latter revolves once in space against the stars every twenty-seven days. That is the *third* power of three. We never see that axial rotation because Luna always faces us, revolving around Earth in that same period. These powers-or-three numbers seem to be required for the (no doubt) vitally important business of facing earthwards: one side of the Moon is always shown to us, and one side of Venus likewise always faces earthwards at her closest approaches.

A woman goes through nine mo(o)nths, nine meetings of the Sun and Moon, during pregnancy. These measure out (mensuration, menstruation, from mens, Latin, 'a month') the period of gestation. That would be a time to ponder how Venus revolves nine times more slowly on her own axis in space, than our Moon. No visible light comes from Venus' surface, so there is no point hoping to see what her surface 'really' looks like. Demurely, she has always concealed this, so no-one knew about it, right through history, until just recently!

<u>Quiz question</u>: Can you find the *Octagon of Venus*?

<u>Answer</u>: Take a date, eg your birthday, and then mark out Venus' heliocentric position in the zodiac at yearly intervals, over eight years. Join up alternate years to get two squares (Martineau, 2005). I have to use an astrology program to do this, setting it to 'heliocentric ie as seen from the Sun.

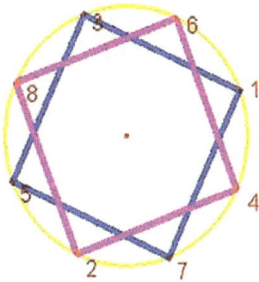

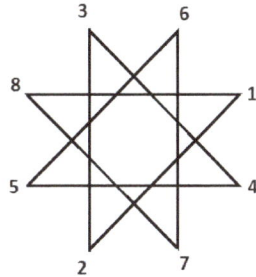

6

Sojourn Across the Sun

Two Venus-transits chimed, in 2004 and 2012. On these very rare events, eight years apart, Venus moves across the face of the Sun. Silhouetted against the Sun, Venus can then appear in the daytime, as a tiny spot that moves across the Sun's face. No living person had ever seen this, when it happened in the summer of 2004.

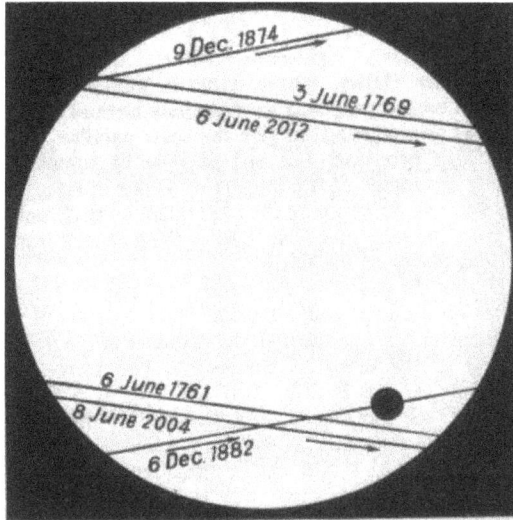

An old transit diagram from *The Flammarion book of Astronomy*

The nodes of Venus are where the plane of her orbit cuts the ecliptic (the ecliptic is the plane containing Earth and Sun). This intersection gives a line in space which is more or less immobile. Whenever a corner of the Venus-pentagram meets this node-axis, a transit happens.

A Venus-transit happens at an inferior conjunction, where Venus has to be near a node, i.e. close to the ecliptic. By way of analogy, eclipses only happen when a Full of New Moon is close to a lunar node.

Transits come in pairs eight years apart and happen at the same time of year around June 8th, also at the opposite end of the year around December 8th as the Sun again crosses over the Venus-node.

The Venus pentagram revolves in space once per twelve centuries, one-fifth of which is 240 years, but, due to a very slight movement of the Venus-nodes, the figure becomes 243 years. In this interval, four Venus-transits will happen, two in June and two in December. We see from the picture, how transits at 243- year intervals have more or less identical paths across the Sun, more or less on the same days of the year!

Transits have only been seen since the 17th century, after telescopes were invented and Kepler's astronomy was published. This was the first time that observations and theories were accurate enough to discover the planetary nodes. In Paris in 1631, Pierre Gassendi was the first person to witness a transit of Mercury, then in Britain in 1639 the young Jeremiah Horrocks cleverly predicted and saw the first-ever Venus-transit, somewhere outside Manchester.

The tiny size of these planets against the Sun staggered everyone. People for the first time could directly apprehend the vast scale of the solar system. Horrocks asked his friend William Crabtree to watch this Venus-transit, and recorded his friend's thrill:

rap't in contemplation he stood, motionless, scarce trusting his senses, through excess of joy.

Crabtree and Horrocks both died young, but their work, which only just and partially survived the ravages of the English civil war - kick-started British astronomy. Horrocks' account kept breaking out into verse. On future transits of Venus, he wrote:

Thy return Posterity shall witness; years must roll

Away, but then at length the splendid sight Again

Shall greet our distant children's eyes.[9]

Venus at Tahiti

In another famous Venusian adventure, Captain Cook set out from Portsmouth harbor in 1768 to observe a Venus-transit on Tahiti predicted for 1769. He discovered beautiful Hawaii on the way, as part of a grueling eight month journey. Captain Cook was a superb navigator; who could find his latitude from the midday Sun, and his longitude from the Moon's position against the stars.

He used an hourglass and a knotted rope to tell his speed. The crew ate sauerkraut, which kept them free from the dreaded scurvy. Only five crewmen had been lost when the ship rounded the stormy Cape Horn. About one in five of his crew got flogged,

H.M.S. Endeavor on the High Seas and Captain Cook.

---

[9] I did two accounts of this: www.dioi.org/kn/venustransit.htm and www.dioi.org/kn/IAUVenus-Transit.pdf

which was quite normal in those days.

Upon finally reaching Tahiti, the ship's young naturalist Joseph Banks told his diary that it was 'the truest picture of an arcadia ... that the imagination can form.'

On the day of the Venus-transit, the king of the island, called Tarróa, plus his sister, Nuna, joined them for breakfast. Later that day, Banks' diary records, there was a visit by 'three handsome women.' He had little to add about the observation of the Venus-transit, which was rather out- enchanted by Tahiti.

Although the French and British had been at war, the French allowed Cook's ship to sail safely, guaranteeing it a safe passage, because they believed he was on a peaceful scientific mission. He sort of was, but there was more to it than that, because, after witnessing the transit, he opened a sealed envelope from the Admiralty, instructing him to explore the South Atlantic. He went on to discover New Zealand, and his expeditions inflicted severe damage upon the indigenous peoples of the Pacific. It's only fair to have a Maori view here:

Venus luring these Europeans
    for a glimpse of her in a glass,
To confirm an astronomical chart
    and the psyche's template
 The starry temple of woman to our crewmates

 The Captain speaks, heading for Tahiti.
    Ah Paradiso bliss!
These natives are the gentlest, a breezy caress
    while my crew
Feast on all their senses can thrust and chew

These friendly islanders don't give a toss.

Haul the anchor, first mate! and drop it again,
It's almost too much with Venus on the brain.
    Rule Endeavor! Endeavor
 Rules the waves, and soon the heavens!

Onwards for Venus! By King George
    it's hard to sleep on the voyage,
Especially when every glittering light
    could be Her skirting across the infinite night
Taking my admiralty with her. Oh to sleep ... [10]

Transits across the Venus node-axis

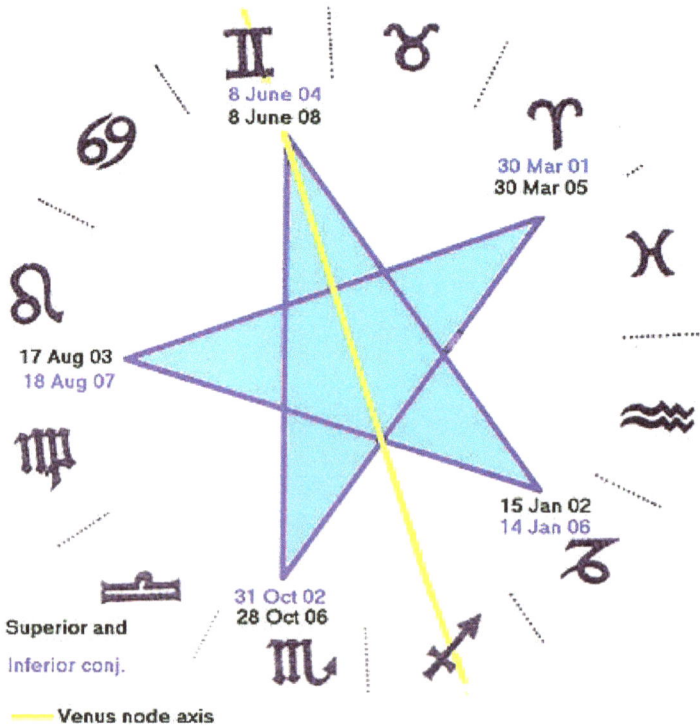

♊
8 June 04
8 June 08

♉

♋

♈
30 Mar 01
30 Mar 05

♓

♌

17 Aug 03
18 Aug 07

♒

♍

15 Jan 02
14 Jan 06
♑

♎

Superior and
Inferior conj.

31 Oct 02
28 Oct 06
♏

♐

— Venus node axis

[10] Robert Sullivan, Captain Cook in the Underworld, 2002, Auckland U. Press.

Having seen two Venus-transits during a period of eight years, we now figure out how they work. The diagram shows how the sky-pentagrams are woven every eight years, for 'inferior' (Sun-Venus-Earth) and for 'superior' (Venus-Sun-Earth) conjunctions. Amazing don't you think? It's helpful to compare this with the diagram at the start of Chapter 3.

Look at the conjunctions in Gemini, they are on the node-axis of Venus, which as we've explained is where Earth's orbit cuts the plane of Venus' orbit. The two orbit-planes are tilted by three degrees to each other, and that's where they intersect. A transit therefore moved across the face of the Sun, on 8th June 2004. Of distinctly less interest, is the fact that four years later on the 8th June 2008 there was an 'occultation', as Venus passed directly behind the Sun, an invisible event

This is very similar to how a lunar eclipse can happen two weeks on either side of a solar eclipse. These events happen every eight years, and they carry on in eight-year chains that span at least half a century, with the two Venus-transits in the middle of this long chain. There was another occultation in 2016, then in 2024 etc., four and eight years apart. Due to perspective and parallax it's quite easy for Venus to go behind the Sun, whereas it needs some exact positioning to appear in front, at the inferior conjunction. One of these events takes far longer than the other, owing to the different directions in which they are moving: the occultation takes two days in contrast with the mere five hours of the transit.

As the pentagram slowly revolves, another pair of transits will happen at the other end of the Venus-node, some 120 years later. If two pairs of Venus- transits turn up every 243 years, does that period ring a bell? It should do, it's our friend the axial rotation period. So we have a day-for-a-year concordance going on here. How close is it?

Venus transit cycle = 243.00 Earth-years.

Venus axial rotation period = 243.018 Earth-days.

There are 365.24 days in a year here on Earth. Venus spins 365.23 times in its transit-cycle, a bizarre agreement to within several parts per million. And we saw how 243 isn't just any number, its the fifth power of three, or 3 x 3 x 3 x 3 x 3. So what does all this mean? It must mean something! We can express this mystic concordance in days:

> *There are 365.24 days in a year and Venus spins 365.23 times in space per transit-cycle.*

Give yourself a year or two to mull over these things - don't try to take it in all at once. The most up-to-date radar orbital satellite data are here required, to ascertain the incredible precision with which Venus and Earth are a-dancing together.[11]

These high-precision synchronies cleverly woven by Venus are more or less as exact as those attained by Luna: the nineteen-year Metonic cycle and the eighteen-year Saros cycle are even more exact than anything that Venus can manage, these two are within literally a few parts per million.

What these relationships have in common, is that there is absolutely no reason (that anyone can discern) why they should exist. They just happen to be there, in the cosmic machinery. They're a part of the heavenly music. Both of the 18-19 year lunar periods have been of central importance for astronomy, one for the preparing of calendars and the other for predicting eclipses; and yet they are both 'mere' coincidences! Only the two traditionally 'feminine' planets, the Moon and Venus, come up with these harmonies.

---

[11] NASA gives the transit cycle as 88,756 days = 243.00 years.

The number five has kept turning up: in the great sky-pentagrams, then a five-to-one Venus-Earth day ratio in the last chapter, then the sequence of five (or so) transits/occultations happening together, and finally a fifth power of the number three. Do they point towards some *quintessence* of Venus?

A French medallion commemorating the 1874 transit of Venus. See her close to the Sun-god Apollo on his flying chariot, while the figure of *Astronomia* on the ground looks up in awe.

The second of two recent Venus-transits happened in the summer of the year 2012, a year for which great things were expected. It chimed on June 5th as compared to June 8th for 2004. A couple of

49

months later London's Olympics were held, echoing the eight-year period of Venus in its ancient tradition.

2012 Venus transit (van!)

Jungle Calendar Reaches Fifth Sun

A calendar started twenty-five centuries ago in the jungles of Peru by an unknown race - one strangely equipped to count through millennia - clicked back to zero in 2012. The Mayan Calendar Round spins on and on, but its Long Count had an end-date. The blood-drenched culture of central America which produced this, was horribly involved in the synodic Venus-cycle. This book is about the Path of Beauty, but our chapter on the subject (Chapter Ten) is the twist of horror that makes it complete! Northern-hemisphere Venus deities were all female (although NB Lucifer as the Roman term for the Morning Star was taken over by Christians to be their arch-fiend, as we see in the next chapter) – but that of Central America was male.

7

The Evening & Morning Star

"Venus is not to be seen at all times, and to those who are not acquainted with her movement she seems to come and go as she pleases. For months altogether the Star of Evening is hidden from mortal eyes. But every movement of the seemingly capricious planet is known to those who study the almanacks. Each step of the Queen of Beauty is given with prosaic detail as she moves along her path, but to those who do not pay much attention to astronomy there is undoubtedly a charm in the way she suddenly makes her appearance as the leading lady in the celestial drama.

"It is a beautiful clear evening, the Sun has just set, and in the golden glory of the western sky a beauteous gem is seen to glitter. A few weeks later the Queen of Beauty has risen higher above the horizon and rides, an even more brilliant object in the sky, long after the shades of night have descended. She only occasionally attains her full splendor, but at such times she outshines even Sirius more than twenty times. Then again she draws near the Sun and remains lost to view for many months, until she enters upon a new cycle of changes after an interval of a year and seven months."[12]

So wrote the American astronomer Mary Proctor, in 1928. Today, anyone fortunate enough to dwell away from the city's neon glare

---

[12] Mary Proctor, *The Book of the Heavens,* 1928 p.98.

may like to note the dates when they can first espy Venus in the evening or morning sky, and then when, months later, she fades away. Supposedly there are 263 days between these, or that is the figure given in old astronomy books. That may not be an exact figure, but roughly it is the average period of human gestation, from conception to birth. The *Rose-and-Heart mandala* that we just looked at does *not* indicate when Venus appears and disappears, but rather maps out its as it were perfect blueprint of sidereal motion, sidera being the Greek word for stars, i.e. Venus' motion against the stars. This chapter is unconcerned with Venus' position against the stars, but looks at how we can experience it against our local horizon: this chapter is more *experiential.*

The Greeks of Homer's day hadn't twigged that the Evening Star, which they called Hesperus, and the bright star of the morning, which they called Phosphorus, were one and the same (four centuries later they made the connection). The Romans likewise used two different words, Lucifer and Vespers, for Venus, as the Morning and Evening star.

The diagram here depicts the synodic cycle as a circle, as it cycles through time. Venus' period of maximum brilliance arrives a month or so after she has risen to her highest in the sky, and several weeks before she swiftly fades away. This is the perfect time for that enchanting cocktail-party you were meaning to organize. Sitting out on the grass or on the balcony, and viewing Venus together as she slowly sinks, is a beautiful social experience, but also one where important issues of philosophical meaning may be apprehended.. Plan ahead for it - or, if that week isn't convenient, turn to Appendix 3 and find a time when the thin crescent moon (waning or waxing) conjuncts the Morning or Evening Star. It's a lovely sight in the sky - hang on, make that the loveliest sight in the night sky.

In antiquity the character of the Morning Star was more like Nike the goddess of war, as if she were more brave and bold when appearing in advance of the Sun. Boldly glittering as herald of the dawn, she caused the stars to fade away. She had the connotation of getting up and striding forth into the world - whereas Aphrodite the seductive love- goddess was more associated with the Evening Star, whose sinking down into the sunset put people more in mind of going to bed at night. Her first appearance as Morning Star (called, a 'heliacal rising') was honored as a good omen.

When Venus re-emerges from the other side of the Sun as Lucifer, bright Star of the Morning, she is still 'retrograde,' although we don't normally notice that in the sky. She is then moving most slowly in the sky as we saw in previous chapters, backwards against the stars. She is moving most slowly against the zodiac at her disappearance and reappearance, as well as moving backwards. This retrograde motion lasts for forty days, while she backtracks over 15-18° of an arc. So there is a complicated sequence to her celestial ballet.

After her reappearance she then becomes stationary, i.e. stops moving backwards in the zodiac, then starts to move forward, becoming most brilliant, and then climbs highest into the pre-dawn sky. It's sheer poetry, really. Radical feminist groups ought to assemble every 19 months, at Venus' maximal brilliance as Evening Star, or maybe in a more 'butch' and strident fashion when the Morning Star appears. That had has been regarded amongst the Mayan people as the right time to start a war.

Returning to the period of gestation mentioned earlier, this also closely relates to nine cycles of the Moon. Nine times do the Sun and Moon meet in the sky from conception to birth, giving us 266 days, and Venus is visible for on average about 263 days. A fairly exact mean period of human gestation can nowadays be obtained from artificial insemination studies, and these seem to give a few days less than 266 days. Thus, amazingly, the two traditionally feminine planets harmonise with the duration of human pregnancy. This was not always understood: in the nineteenth century, nurses reckoned that pregnancy lasted forty weeks (280 days). That was way off! That extra-long value alluded to the last period a woman recalled having had, before conception took place, as being the event that could be dated.

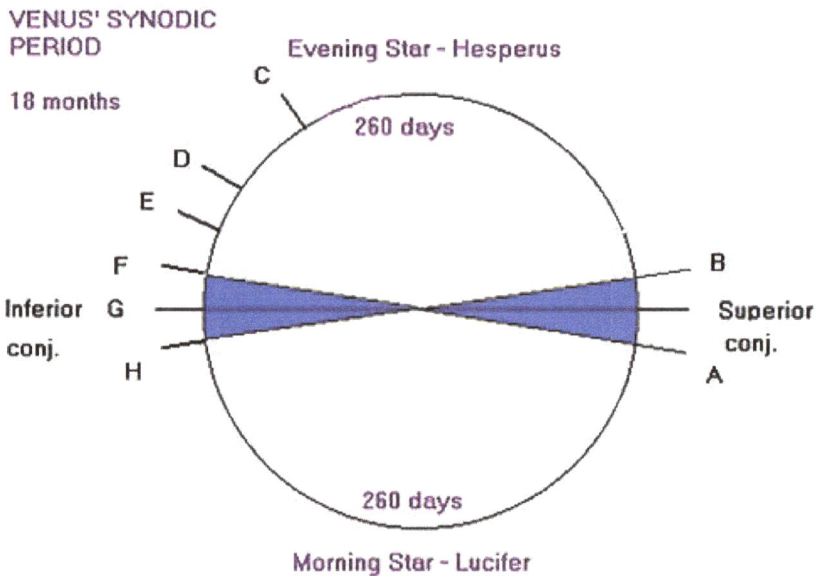

VENUS' SYNODIC PERIOD

18 months

Evening Star - Hesperus

C

260 days

D

E

F

Inferior   G

conj.

H

B

Superior

conj.

A

260 days

Morning Star - Lucifer

The synodic or experiential cycle of Venus, where A–B is roughly 40 days, C–D is 36 days, D–E and E–F are 14 days, and F–G is 8 days

Impress your friends by predicting some of these cycles, starting off with the dates of solar conjunctions (Chapter 3). Subtracting six weeks from an inferior conjunction date gets the week or so of maximal brightness of the Evening Star. Check your predictions against a 'Venus diary' on the web which gives relevant dates, year by year.[13]

Venus has a longer synodic cycle than any other planet.

Lucifer: From Morning Star to Arch-Fiend

The first bible texts in Greek were called the 'Septuagint,' composed in the third and second centuries BC somewhat as a translation of old Hebrew sacred texts. It was made in the city of Alexandria, probably at its newly-established library. Its Book of Isaiah (14:12) has the immortal line,

> How thou art fallen from Heaven O Lucifer, Son of the Morning!

- at least, that is the King James' version of it, with which we are familiar, where 'Lucifer' is a Latin translation of the Greek 'phosphorus.' The old, Greek word for the Morning Star was 'phosphorus.' Both of these words enjoyed the dual meanings of 'light-bearer' and 'Venus, the Morning Star'.

Isaiah's original Hebrew text from which this Septuagint translation was made – inscribed on a leather scroll, from the 8th or 7th century BC, a copy of which was found amongst the Dead Sea Scrolls - goes:

> How have you fallen from heaven O Helel, son of Shahar!
> How art thou cut down to the ground, which didst weaken the nations.

'Helel' and 'Shahar' being obscure Canaanite deities. Shahar was a dawn-god, so does that mean that his son would be Venus, the Morning Star? Isaiah was having a tirade against the King of Babylon, who would, he affirmed, soon be descending

---

[13] www.astro.com/swisseph/ae/venus1999.pdf

into Hades. To be sure, the text sounds a hell of a lot more exciting using Phosphorus/Lucifer rather than 'Helel'.

The lovely meaning of what the Roman empire knew as their Venus god 'Lucifer' had been extolled by poets:

> Aurora watchful in the reddening dawn, threw wide her crimson doors and rose-filled halls; the stars took flight, in marshaled order set by Lucifer who left his station last. (Ovid, Metamorphosis, 2.112)

> And now Aurora, rising from her Mygdonian resting-place, had scattered the cold shadows from the high heaven, and, shaking the dew-drops from her hair, blushed deep in the sun's pursuing beams; toward her through the clouds, rosy Lucifer turns his late fires, and with slow steed leaves an alien world, until the fiery father's orb be full replenished and he forbid his sister to usurp his rays. (Statius, Thebaid, 2.134)

Later on in the New Testament, the Prince of Peace was praised by a comparison with the Morning Star:

> I, Jesus have sent my angel to give you this testimony for the churches. I am the root and the offspring of David, and the bright Morning Star.

That is from the Book of Revelation, written about AD 95. Finally, in what experts believe was the last book of the New Testament to be composed, we find

> *until that day dawns and the Morning Star [Phosphorus] rises in your hearts.*

-Second Epistle of Peter

There are three references to the Morning Star in the New Testament, alluding to Jesus or to the Holy Spirit. The Morning Star 'is a consistently positive image and one that is solidly associated with Jesus the Messiah.'[14] So there!

How did it come to pass that, once the Old and New Testaments were bound together, they had the Greek word, 'Phosphorus' alluding both to the Savior, Jesus, and to a being who was soon due to be dramatically re-cast as the Lord of Hell?

Fall of Lucifer: Gustav Doré's illustration to Milton's *Paradise Lost*

Origen of Alexandria (185 – c.250) wove together the story of Lucifer and the fallen angels cast from heaven. Then, in the fourth century the 'Vulgate,' a Latin translation of the Bible by Jerome, became Europe's standard Bible for over a thousand years, and it

---

[14] *Satan a Biography* by Henry Kelly, 2006, p.167.

used the word 'Lucifer' for the Isaiah passage. Modern Bible translations seem not to use the word Lucifer in the Isaiah text, as if not willing to accept the Origen demonology.

Strike a match, and watch its phosphorus burn. From Dante to Jagger, what could be a mere translation error has triggered off the awesome image of a fallen Ruler of the Underworld.

8

The Martineau Symmetries

As the geometer who sets himself
To square the circle and who cannot find,
For all his thought, the principle he needs,

Dante *Paradiso* Canto 33

For a thousand years – no, make that two thousand years – astronomers and philosophers in the West believed that circles and spheres belonged in the heavens. The explanation of how things moved up there, used circles. This was because they were viewed as being somehow divine and perfect. I guess Plato started it. But, that was all over with centuries ago, wasn't it?

In Chapter Two we looked at a Mercury-diagram by John Martineau, based on three circles. We here come onto his work with Venus, but before doing that let's go back to an earlier construction by John Mitchell, who had been a friend and mentor to the young

Mr Martineau. This famous construction concerns the Earth and Moon, and the idea of squaring the circle. Such ideas must bring us to the question of perfection: could the system be somehow *perfect*?

Looking at this early case-study, may help you to decide, whether or not you believe there is any value in this 'ideal' circular perspective.

John Michell imagined bringing the Moon close, until it just touches the earth. He then noticed, that its centre would go through a circle whose circumference has the same length, as the sides of a square, which contains the Earth.

Let's first of all agree, that such a concept has a decorative and aesthetic and perhaps an architectural meaning, quite apart from whether it has any astronomical significance.

What is called 'squaring the circle' was for a thousand years regarded as the most baffling enigma. For a circle to have the same perimeter as a square, the ratio of its radius to the half-length of one of the sides of the square, will be $4/\pi$.

In the above diagram, there are two concentric radii of the large and smaller circle, the smaller being inside the square. For 'squaring the circle' their ratio would be given by $M+E / E = 4/\pi$, where M and E are the radii of Moon and Earth. These two figures tie up within 0.04% - which is OK.

You might prefer here to use the 22/7 close approximation to pi. The slope of the Great Pyramid tends to come in here! If you imagine the diagram built up starting from the 3-4-5 triangle as shown, then the length of the square enclosing the Earth will be 11 units, just as that enclosing the Moon will be 3 units. So the key ratio is 3/11 and that is a *very* close tie-up.

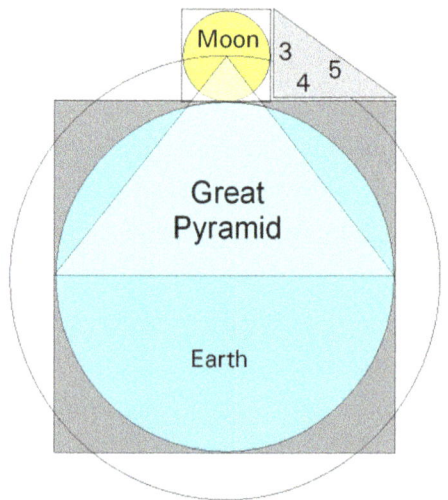

The moon's radius is 27.3% that of Earth, and in fact the exact NASA figures give the size ratio Moon / Earth as 0.2727. Then, 3/11 = 0.2727 – there is no error! Also, the greater circle in our diagram will be 1.2727 that of Earth, where for comparison, $4/\pi = 1.2732$. The circle is squared!

It's hard to resist the image of the Great Geometer up above at this point! There will be one more of these ratios involving phi, that we come to later on.

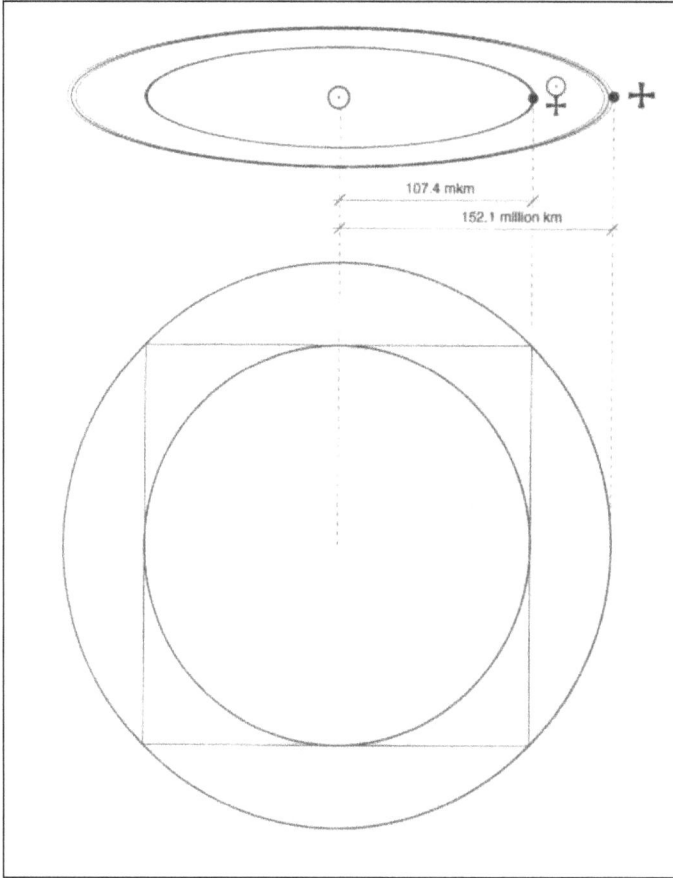

107.4 mkm

152.1 million km

The 'Realm' of Earth and Venus

That is enough for now, to give a brief perspective, on whether or not circles have got any business up in the sky. Astronomers would say that the rings of Saturn are very close to being perfect circles, but that isn't quite what we are dealing with here, we seek for *ideal* meaning. Also the diagrams we've looked at here, are something you might want to call, or that a lot of people do call, sacred geometry.

Here's another 'square' design. It looks at the near-circular paths of Earth and Venus around the Sun, from Earth's greatest expansion (called, *aphelion*, i.e furthest from the Sun) to Venus' perihelion, i.e. it considers the slight breathing expansions and contractions of their orbits. Here's how Mr Martineau expressed the matter:

> *Marital Bliss - In which the realm of Earth & Venus is taken as a single space with a simple geometry: Earth and Venus are so very happy together that it is their whole combined space which exhibits the simplest harmony. Between Earth's greatest distance from the Sun and Venus' closest approach to the Sun lies the total realm of Earth & Venus, their home. A single square proportions this region with 99.9% accuracy. The square was generally associated with Earth, the City or the Home...*

Earth reaches its greatest distance from the Sun each year in July: around the time of greatest heat, then we have drawn furthest away from the Sun. We earlier looked at how 'synodos' means meeting, and how the cycle of meeting and separating can be experienced as Venus appearing and then fading away from our view, as the Morning and Evening star. [15] The diagram here shows that moment of meeting, when two spheres draw closest together in their paths around the Sun. They meet together over that synodic period.

---

[15] The ratio of the two circles in this diagram is as the square root of two: to check the maths, look up the Venus perihelion (nearest to the sun) and Earth aphelion (furthest) distances on NASA pages, and find their ratio. It will match the square root of two to 0.07%.

The next Martineau symmetry we come to, reaches a higher order of precision. Putting eight circles touching each other around the Venus-orbit, gives Earth's orbit, to an awesome 99.99%, or one part in ten thousand. We here use the mean orbits which are perfect circles. The design is eightfold, that being a fundamental number for the Venus-Earth music. This design in Space echoes what we previously experienced in Time.

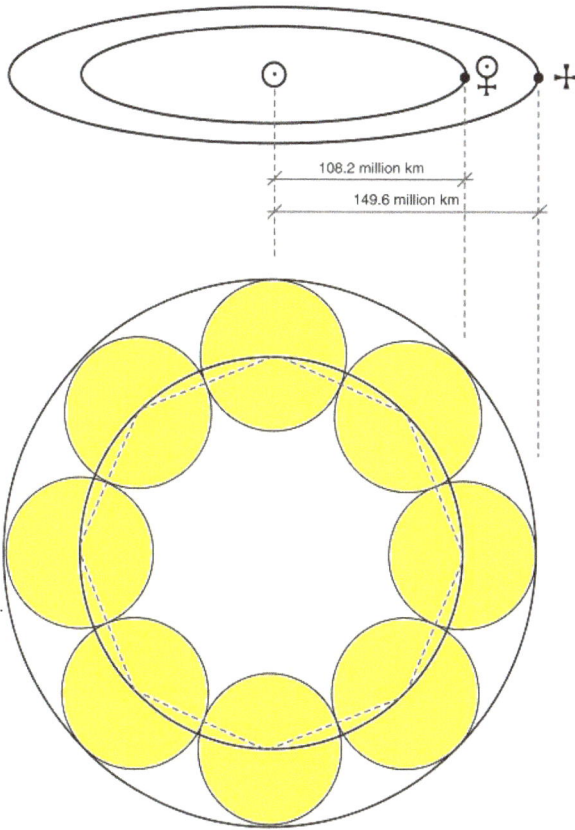

Eight touching circles define the mean orbits of Earth and Venus, to one part in ten thousand

So, 'What does it mean?' you ask? Mean? Well, it may help to notice something about the number of days in Venus' synodic cycle:

$$8 + 8^2 + 8^3 = 584$$

So eight, plus eight squared, plus eight cubed, gives its synodic period in days. The period of 584 days it isn't just any old number, it's the sum of powers of eight. Well, fancy that! This eightfold pattern can also be compared with the ancient symbol for Venus: an eightfold star, which we come to in the next chapter.

John originally claimed a 99.9% accuracy for this, in his now totally forgotten *Book of Coincidence* of 1995 where he described these things. I obtained the most exact planetary orbit data then available - from the Royal Astronomical Society's library - and was startled to ascertain (using the formula Earth/Venus = $1 + \sin \pi/8$) that it actually had one whole order of magnitude higher accuracy, i.e. 99.99%.[16] It was my re-calibration of his symmetries that drew me into this subject. John had not used the algebraic formula above, but some other geometric method. Indeed the appeal of this book came from the absence of any algebraic formulae, it was purely geometrical. He refused to reprint this great masterpiece, only ever publishing a thousand copies - for some aristocratic reason - and it's now long forgotten.[17] I never saw a book which had more effect upon people.

Venus, Earth and a Pentagram

Again we look at the mean orbits of Earth and Venus, and this time the circle just touching the pentagram is that of Venus. The inner circle of the pentagram, is related to the outer one, as the *square* of the Golden Ratio. In Chapter Four we found that the square value of ϕ is $\phi + 1$, i.e., 2.618. Give yourself a year or two to

---

[16] Martineau, *The Little Book of Coincidence* 2000.
[17] But, he kindly allowed me to put some of it up on the Web: www.astrozero.co.uk/astronomy/planets.htm

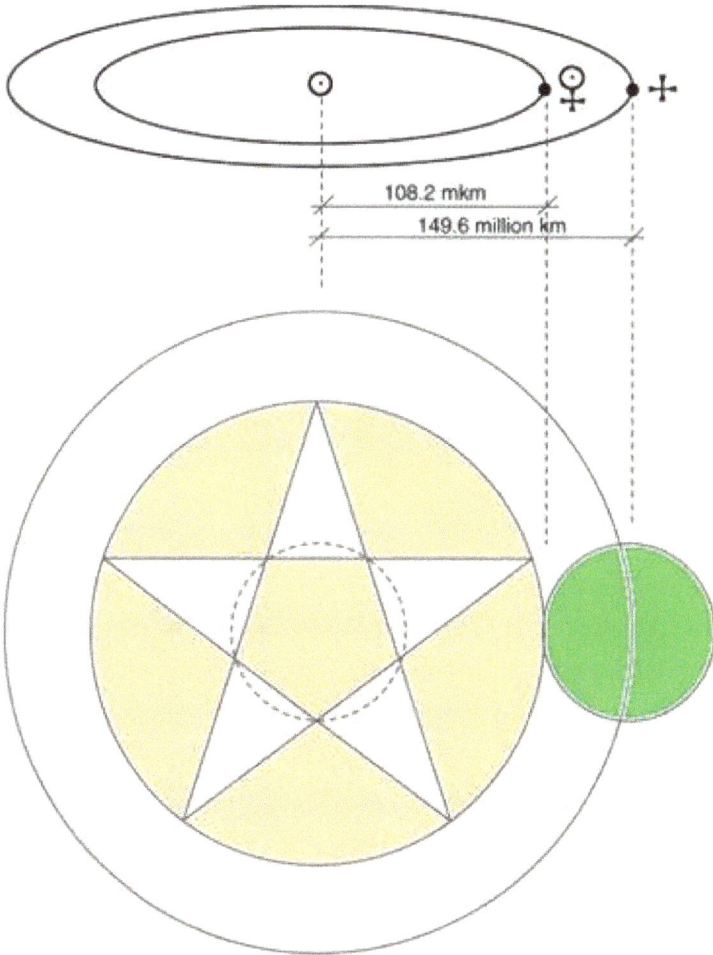

A pentagram fits Earth and Venus

mull over this pentagram symmetry. It works to 99.9%.

The 25-year old John Martineau discovered this construction in 1992. In Chapter Four we saw how the Golden Ratio linked the years of Earth and Venus together, whereas here that same ratio is appearing in their orbit distances.

Notice how the diagram does not say anything about φ, it just invites appreciation in a visual manner, without any maths: it's a picture, of two mean orbits and a pentagram. One would like to see a Venus-temple having a marble floor-design, of one of these patterns.

Coming back to John Mitchell's construction that we started off with, it will work just as well with phi instead of pi. How can that be? The right-angled triangle here shown has sides phi, square root of phi and one. That triangle gives the slope angle of the Great Pyramid! The base angle here given agrees with the pi-triangle given at the start of this chapter, to within a single arcminute. Thus, pi and phi here come together. Dividing the two sides of the triangle, as we did before, gives the equation for the Earth and Moon radii ratio, as $\sqrt{\phi} = (E + M)/E$. The square root of phi is 1.2720 and this defines the Moon-Earth size ratio to one part in two thousand! That is precision! I'm not sure if John Mitchell ever realized that his argument applied to both pi and phi, or if so he kept quiet about it.

That the Moon-Earth system is doing this, mathematically indicates how closely connected it is to Mother Earth.

Applying Pythagoras' theorem to the triangle in the diagram, we get $\phi^2 = \phi + 1$: surely the first use of Pythagoras' theorem ever, it defines the slope angle of the Great Pyramid.

What concerns us here is the idea of sacred geometry, and that refers to a structure that has a deep symmetry and is pleasing, i.e. it works at different levels of meaning.

Early on, John Martineau looked at Mercury, and how its relative size compared with Earth was exactly held by a pentagram. Again we have a ϕ^2 ratio, and using NASA's values of the mean radii of Mercury and Earth, their ratio is 0.3829: which is very closely $1/\phi^2$.

He noticed that a regular octagon gave a very similar ratio to that pentagram construction, here they both are:

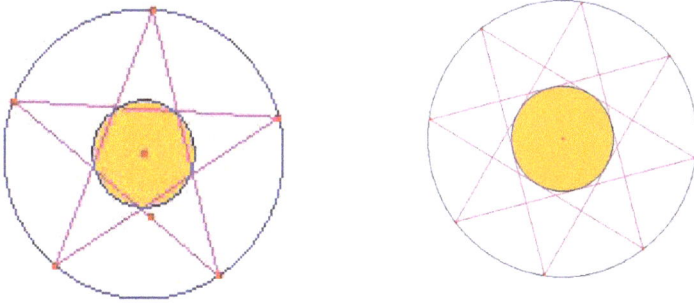

Earth and Mercury mean orbits

Just for fun, let's write the equations here, in terms of pi and phi. The ratio between inner and outer circles is here given by:

Octagon: sin $(\pi/8)$ = 0.3826   Pentagon $1/\phi^2$ = 0.3820

- you can see they are pretty close. The Earth-Mercury size ratio fits the octagon pattern to 0.05% - as likewise the *mean orbits* of Earth and Mercury also fit into these same patterns, to within 1%. That is not quite so exact, but still quite impressive.

Are such matters 'mere' coincidences? They might help to inspire architects or light up a dreary school maths lesson. HRH Prince Charles' book *Harmony* aimed to inspire architects and town planners and it featured some of these Martineau Symmetries. There were about seventy in his book, now sadly forgotten, of which we've here covered half a dozen.

These high- precision symmetries may remind us of the words of Johannes Kepler in his *Harmonices Mundi* of 1618, about the harmonies of the world:

Now there is need, Urania, of a grander sound,

while I ascend by the harmonic stair of the celestial motions to higher things,

67

where the true archetype of the fabric of the world is laid up and preserved. [18]

The great astronomer then added:

If anyone expresses more closely the heavenly music described in this work, to him Clio pledges a wreath, Urania pledges Venus as his bride.

Whoa, steady on there! But, there are rewards for studying the celestial harmonies.

---

[18] Opening of Chapter 7 of Book 5 of his grand master-work, English translation *The Harmony of the World by Johannes Kepler* 1997, translated by Eric.Aiton *et al.*

9

From Ishtar to Aphrodite

At the end of the day, the Radiant Star,
The great Light that fills the sky,
The Lady of the Evening appears in the heavens,
The people in all the lands lift their eyes to her.

Sumerian poem to Inanna

Venerate and *venereal* derive from the same root, and following their etymology takes us a long way back in time, to a love-goddess. Looking up to somebody with *reverence* means that we venerate them - not so easy these days - whereas *venereal* issues signify trouble down there where a visit to the hospital may be called for.

These two words are so very opposite in their meaning, the one pertaining to spirit and the other to body, that one is hard pressed to imagine a single source for them. We here try to do this, by recalling some ancient deities. These were linked to the eight-pointed star of Venus: Inanna! Ishtar! Aphrodite! Worshiped and adored through millennia, the first two were love-and-war goddesses, while Aphrodite, though she had some reputation for assistance in war, was really *the* love-goddess. Other planets had male and female deities, but only Venus was solely feminine.

Inanna

Inanna of ancient Sumer was a dominant goddess who ruled. When the Babylonians arrived they continued to worship her but called her Ishtar instead, and She was a bit more militaristic. Like Aphrodite much later on, Inanna would couple now and then with a mortal, so that person became king: she chose kings and this did involve ceremonially lying with them. Similar comments could be made about Ishtar: 'But thou, O Ishtar, fearsome mistress of the gods, Thou didst single me out with the glance of thine eyes.' The king acquired the powers of kingship from making love to Inanna.

According to the Sumerian myth, the Huluppu tree was planted beside the Euphrates 'in the first days when everything needed was brought into being.' But the South Wind uprooted it, and the waters of the Euphrates carried it away. Inanna appears, as one 'who walked in fear of the sky god, An,' not yet a queen because she lacked the emblems of her divine status. She rescued the tree and planted it in her garden in Uruk. But, pests invaded her tree: a snake 'who could not be charmed' made its home in the roots, and an anzu bird (part eagle and part lion) reared its young in the branches, and Lilith, a maid who was part woman and part owl, built her home in its trunk. Inanna tried to get these parasites out of her tree, but couldn't. She wept and called upon her brother the Sun, Utu, but he just turned his back on her. Then she called to Gilgamesh, who was the hero of Uruk. He arrived wearing sixty pounds of armor and a large bronze axe, with which he slew the serpent, whereupon the anzu bird and Lilith both fled. The tree became part of the institution of sacred kingship for this earliest literate civilization, connecting the gods in heaven and their kingdom on earth.

Another story tells how the main city of Uruk acquired its wisdom. Inanna decided to pay a visit to Enki, the god of wisdom: 'I shall honor Enki, the god of wisdom, in Eridu' she said, and when she got there they had a drink together. Rather inebriated, Enki rashly shared the various decrees called *me* to Inanna which were the principles of wisdom, on all sorts of topics. When he had sobered up he regretted this, and sent various monsters to attack the departing Inanna in her Boat of Heaven. She however had summoned a warrior goddess companion, Ninshubur, who was able to protect her. When they arrived in Uruk the people lined the canal rejoicing, and the *me* were given out amidst the sounds of drums and tambourines. After Enki has given the *me* to Inanna, he then realises the next day that he doesn't have them any more! We might wish to translate *me* as technical know-how, but it weirdly sounds more like, original software where one lacks a backup copy!

On another occasion, the celestial deity Inanna descended into the underworld. She wanted to attend the funeral of Gugalanna, the Bull of Heaven, who was the husband of Ereshkigal. Although a sky-goddess, she could make that descent, because of a gift of power she had received from Enki. She prepares for this by dressing up, for example she puts on some seductive eye-shadow called 'Let him come.' She then passes through the seven gates of the Underworld, being obliged to remove one of her *me*, i.e. her items of sovereignty and divinity – her crown, her royal robe, her measuring rod and so on, at each step.

Finally, she arrives naked before the Queen of the Dead, much-feared Erishkigal; who kills her without a moment's hesitation, and hangs her corpse on a hook. It hangs there as a carcass for three days, but meanwhile Ninshubur (the protector of Inanna) laments, and seeks for aid. The deities Enlil and Nanna were indifferent, but Enki agrees to assist. He fashions two androgynous creatures, who are able to pass through the underworld planes. They take with them some nourishing medicine, that is able to revive Inanna, and descend. Upon arrival, they express empathy and support for Erishkigal's troubles (her husband has just died) and thereby extract from her the promise of a favor. They beg to be allowed to take Inanna's corpse, which they then restore to life, and bring her back.

A boon has to be paid, however, because the law of the Underworld demands that some substitute be sent down. Inanna's husband Dumuzi has to go instead, on account of the way he failed to lament when he heard of her death! His sentence was commuted after he bewailed his fate, so he was only obliged to go down into the underworld for some months every year.

Poems of Enheduanna

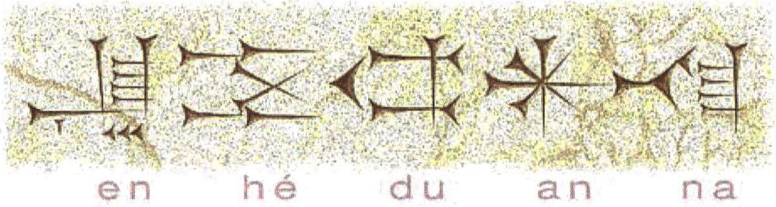

Figure: name of the poetess, in Sumerian writing

Inanna turns up in the First Poem, way back in the 3rd millennium BC - when no nation except Sumeria had writing. If you thought Sappho was the earliest poetess, think again! *En-he-du-an-na* was high Priestess of *Inanna* at Ur.

As the first author to sign a text and the earliest author known by name, who wrote in the first person, she was the daughter of Sargon, king of Sumer, and lived around 2300 – 2225 BCE. Here are two excerpts from her long incantation to Inanna. They may remind us of the earlier story, about the *Me*. The second part has something a bit like a Moon-Venus conjunction. We rejoice to hear its craggy, primordial tone:

> The Great queen of queens,
>
> born for the rightful ME, born of a fate-laden body,
>
> you are even greater than your own mother,

full of wisdom, foresight,

queen over all lands who allows existence to many,

I now strike up your fate-determining song!

All powerful divinity, suitable for the ME,

that which you have said magnificently

is the most powerful!

Of unfathomable heart,oh highly driven woman,

of radiant heart, your ME,

I will list for you now!

Into my fate-determining Gipar, I had entered for you.

I, the en-priestess,

 I, En-hedu-Ana.

While I carried the basket,

I struck up the song of jubilation,

as though I had not lived there,

they offered the death sacrifice.

I came close to the light,

there the light became scorching to me.

I came close to the shadow,

there it was veiled by a storm.

My sweet mouth became venomous.

That with which I gave delight, turned to dust...

The Queen, the strong one,

the ruler over the gathering of the 'en',

she did accept her prayer and sacrifice.

The heart of fate-determining Inanna has turned to its place.

The light was sweet for her,

delight was spread over her,

full of abundant beauty was she.

As the light of the rising moon,

she too was clothed in enchantment.

Nanna came out to rightfully gaze in awe,

He and her mother Ningal blessed her,

And then the gate post said unto her "Be hailed!"

What each said to the nugig is exalted.

Destroyer of enemy lands,

endowed with the ME from An,

My Queen, draped in enchantment,

to you Inanna be glory!

Enheduanna received her name when ordained as a priestess of the Moon (from *en,* high priestess, *hedu,* ornament, *ana* of the sky-god). Her primal poem, her jubilant song, was to Inanna. It's hard to believe these are words from 4,300 years ago, when pyramids were being built. The hundred or so tablets recording her incantations date from a period five centuries later, but are reckoned to be copies of earlier material. Decent English translations appeared only recently, but hardly anyone seems to be interested.

Here for comparison is a hymn to Ishtar, in the later Babylonian culture of c. 1600 BC:

She is clothed with pleasure and love.

She is laden with vitality,

charm and voluptuousness.

Ishtar is clothed with pleasure and love.

She is laden with vitality, charm and voluptuousness.

In lips she is sweet; life is in her mouth.

At her appearance rejoicing becomes full.

She is glorious; veils are thrown over her head.

Her figure is beautiful; her eyes are brilliant.

The goddess - with her there is counsel.

The fate of everything she holds in her hand.

At her glance there is created joy ..."

12th century BCE Babylonian Kassite kudurru of goddess
and priest.

One could go on, but I'll resist. Instead, we turn to a Babylonian
image from the 12th century BC, which could be the earliest
astronomical diagram with a discernible meaning. We can see that
some serious business is going on, as a king presents his daughter
(not shown, but he holds her hand) to the goddess. Hanging in the
sky are three calendrical emblems, of the Sun, Moon and Venus.
Both the Sun and Venus are eightfold so its not immediately

evident, which is which. How low they seem to be, and close to human affairs!

The *Octaeteris*

Two of these emblems are eightfold, why should that be? The eight-pointed star was *the most central and enduring image of Inanna the sky-goddess and Queen of Heaven of the ancient Sumerians* - as worshiped in ancient Mesopotamia for four thousand years: 'The star represents her astral form, the morning and evening star.'[19]

A tablet from the 7th century BCE states that Ishtar of the Evening Star was female, while Ishtar of the morning star was male, so some gender-bending was going on.

The Babylonians recorded the eight-year cycle of Venus in cuneiform tablets, and that star appears as a visible image of this pattern. But, the Solar image (left of the picture) is also eightfold, because the Babylonian calendar-priests knew that the lunar months came into a rough synchrony to the year per eight years. This was a kind of early version of the 19-year Metonic cycle, and it took the form of a Greek myth. The priests were then intercalating - that is, adding in - three extra months: so that every 8 years, three years enjoyed a thirteen month. Thus two major calendar-structures are here shown, whereby the measure of Time is being indicated: that from Venus being established earlier as far as anyone knows than that of the Moon.

The Moon in this image looks as if it really were changing its shape: it grew large, then gradually shrank and vanished away and then reappeared again. Standing on their ziggurats, the astronomer-priests never twigged as did the more philosophical Greeks later on that the Moon was spherical in shape and only

---

[19] 'Inanna the star who became queen', by Beverley Moon, in *Goddesses Who Rule*, ed. Benard & Moon, 2000, p.69.

shone by reflected light, they lived in a more magical world. The Greeks called its 8-year period the octaeteris, and used it in their calendar.

We've seen how, in the 17th century BCE, a Venus-tablet in the library of Nineveh gives a cycle of omens with dates over an eight-year period, the omens being for the first appearance of Venus above the horizon and for the date of its last visibility. There were ten such events per eight years.

A millennium later, the Greeks halved this basic 8-year period, and started measuring out their history by these intervals, called Olympiads. The Olympic games became a pan-Hellenic four-yearly festival, when all wars and feuding had to cease.

Venus returned to the same degree in the sky on the same day over the period. There was no recurrence of the lunar months in their four-year period, but Venus would have chimed in here. If one saw the thin, crescent Moon together with Venus on some special day of the year, then eight years later they would again be together on the same day, in the same parts of their cycles, give or take a day or so (See Appendix 2 to check this up). This calendar comes from an early, myth-making period.

By the Styx the gods swore their binding oath, to break which entailed keeping silence for a year and exile for nine years (eight years by our reckoning): this is a great year, when all the stars and planets were said to return to their original position, a period that recurs in Greek myth and ritual. Mythically it was connected with the idea of being thrown out of heaven: in the Iliad, Hephaestus the smith gets punished by his mother Hera, who cast him out of heaven and he was kept for a great year (eight years) by Thetis. The early Greek astronomers used its 99 moons for their calendar and timing the Olympic games, and did not elaborate on the Venus-cycle that went with it.

Nowadays, the Olympics chime with the leap-year and the US presidential elections, but no-one remembers the eight-year cycle that originally timed it. The Olympic games began as a foot-race over a stade i.e. round a stadium, scheduled by the octaeteris and on the 8th Full Moon of the year i.e. in July or August. There were ninety-nine moons every eight years, which meant that the Olympics had to alternate between fifty and forty-nine moon intervals.

In the year 776 BC the first competitions were held in Olympia in the western Peloponnesia of Greece. Long before this had become the temple city of Zeus it was dedicated to the goddess Ge. Only free Greeks could compete, so long as they had not committed a murder or behaved indecently in a holy place, and they had to run naked, after having trained for ten months beforehand. The competition was a sprint along a stade, a distance of about 200 m. In 720 BC a longer run was added of two stades, and then gradually games were added.

In the Greek story, a beautiful maiden, Atalante, running naked, always managed to outrun her competitors (as a child, she had been nursed in a cave by a she- bear). To win her, suitors had to compete with her in a race and outrun her, but alas they always lost and ended up with their heads impaled, for such was the grisly price for losing. Finally, Venus got fed up with this dire state of affairs and decided to intervene. She gave three golden apples to the suitor Hippomedes. Each time Atalante started to gain, Hippomedes would drop a golden apple, and she stooped to pick it up. Thereby Atalante lost the race, and her chastity. Experts view Atalante as an Artemis-type heroine, Artemis was goddess of the hunt and of the thin lunar crescent. The three golden apples may be solar in their symbolism (gold being the Sun-metal) and represent the months that had to be added in or 'intercalated' every eight years in order to keep the solar and lunar calendars in step. This is a lunar-solar calendar myth, having a slight touch from Venus. This ancient calendar faded in the early centuries BC, to be replaced by the more exact 19-year Metonic cycle.

Aphrodite, the love-goddess

Like Inanna, Aphrodite was represented by an eight -pointed star and associated with both the morning and evening star. The Greeks often referred to her as foreign goddess and as 'the Cyprian', even after she became one of the most important deities in their pantheon. She had come from the idyllic Mediterranean isle of Cyprus, source of copper (latin *Cuprum*, from Kyprus) and the tall, cypress trees. Cyprian wine was famous, and an inscription on a wine-jar from the 6th century BCE reads, 'Be happy and drink well.'

Here She was worshiped from the 12th century BCE in Cyprus, until the Roman empire was Christianised in the 4th century AD.

The blind poet Homer described her as *bitter-sweet, laughter-loving* and *golden-crowned.* Birds were sacred to her, and she had a magic girdle that made the bearer irresistible, by 'the whispered endearments that steal the heart away, even from the thoughtful.'

Images of Aphrodite from Salamis (on the Eastern coast of Cyprus), 4th cent BCE

Her oldest temples were at Paphos on Cyprus, upon whose shore she was born in the myth, whence Bottichelli's image was derived.

Aphrodite provided a royal heritage for the Greeks through the story of her liaison with Anachises, on Mount Ida. She appeared before prince Anachises who was so awed by her beauty that he was convinced she must be a goddess, and offered to build a shrine to her. She persuaded him that she was just a Phrygian princess, brought hither by Hermes to be his wife and have his children. So they lay together and then afterwards she reveals her divinity. Anachises turns away in fear, covering his face with a cloak, but she reassures him. Thus a handsome lineage of Greek kings was initiated in Asia Minor (Turkey).

Did Aphrodite have warrior powers, like Inanna and Ishtar? In Homer's Iliad she is depicted as a goddess who lacks skill in warcraft. Her father Zeus disparagingly advises that she should stick to 'the lovely secrets of marriage' and not involve herself in the Trojan war. And yet, after Hannibal inflicted defeat upon the Romans, they turned to Venus/Aphrodite for assistance: like Inanna and Ishtar she was associated with victory in war. Aphrodisias in southern Turkey has letters and correspondence from Roman emperors with offerings and Julius Caesar wrote to Her there. In ancient Rome, 'Venerari' in Latin was a verb meaning to take an attitude of hospitality, whereby humans sought to attract the benevolence of the gods. The noun coming from it meant graciousness or charm, and that became personified into a Roman goddess, Venus. She was worshipped as a protector of gardens and ruler of the month of April – but lacked any connection with her traditional planet (which they called, Lucifer/Hesperus). Then, during the Punic wars the Roman Venus came to amalgamate with the Greek Aphrodite.

The beach at Paphos, where Aphrodite came ashore?

She walks in beauty like the night
Of cloudless climes and starry skies;
And all that's best of dark and light
Meet in her aspect and her eyes:
Thus mellow'd to that tender light
Which heaven to gaudy day denies.

(Lord Byron)

Quiz Question: When were Olympic Games held in Britain?

Answer: Thrice has Britain hosted the Olympic Games, in 1908, 1948 and 2012. Spaced by the 8-year period of Venus, these three form the 5 : 8 Fibonacci ratio. Their total interval of 104 years is 13 Venus-cycles, which is 'the Great Venus Round' of the Mayans.

Sources for this chapter: 'Goddesses who Rule' Ed Eliz. Benard & Beverly moon, OUP 2000 (Ch. 1 Aphrodite, Ancestor of kings B Moon, Ch 4, Inanna: The Star who became Queen, B.Moon). 'Akkadian Hymn to Ishtar' translated by Ferris J.Stephens. Inanna, Lady of the Largest Heart, Poems of the High Priestess Enheduanna Betty De shong Meador 2001 U of Texas. Sections 60-73 and 143-153 of 'Queen of Countless Divine Powers.' www.angelfire.com/mi/enheduanna   E.C.Krupp, Beyond the Blue Horizon, OUP 1991. Cyprus, the Sweet Land, P.Satavrou, Nicosia, 1971. Greek Mythology John Pinsent 1969 p.25,39

10

Botticelli's Inspiration

The first-ever *exact* use of the Golden Ratio could well have been in Botticelli's *Birth of Venus.* The dimensions of its frame are given exactly, and dividing length by height gives 1.614 - compared to 1.618 as the 'correct' value. That's exact to within one part in five hundred!

Art and Science have here come together. The above figures came from Wikipedia. Then, visiting the Uffizi gallery in Florence, where the picture resides, I checked out in its bookshop another estimate of the dimensions of its frame given in inches - it also gave that same precise ratio.

More recently (2018) I re-checked the Wiki page for this picture, which had earlier given its size as as 172.5 cm by 278.5 cm, and it had an adjusted value for the frame width as 278.9 cm. Does it matter, a few millimetres on the frame of a five-hundred year old picture? Well it brings the phi-value of the picture's frame to well over one in a thousand! Even I am finding it difficult to believe that Botticelli could attain that level of precision. Remember, there was then no mathematical 'value' for phi – that did not appear for at least another century.

Botticelli painted his masterpiece in 1483. He probably knew the mathematical author Luca Pacioli, who later on composed *De divina proportione* ('On the Divine Proportion') in 1509. Pacioli's book gave no value of phi, he gave no fraction or fractions that approximated to it. Sources in antiquity such as Euclid likewise gave no numerical value for it: they did for pi, but not for phi. Pacioli's book was geometrical, it had wonderful drawings by Leonardo da Vinci, and wove an analogy between the Divine Proportion and the Holy Trinity and discerned the phi-proportion in the sides of a pentagon. None of that would have helped Botticelli in scaling up his board, on which the world's favorite picture was to be painted.

His model for the painting was his muse Simonetta Vespucci, born in a fishing-village *Porto Venere,* so named because it had a little temple to Venus from the 1st century BC.

Botticelli would have constructed a square, then drawn an arc as shown reaching from the centre of the base to its top right corner, brought down to the lower right corner of the picture. That made a Golden rectangle. In the final picture, we notice an *elemental quality* of that square-line whereby it divides everything — to its right is solid land and trees, to its left, the sea and sky.

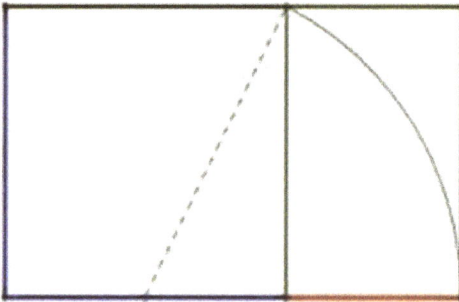

Figure: Making the golden ratio - Botticelli's first step.

He now has a frame built with the exact ratio Φ : 1 – which, to remind ourselves, was given in time by the years of Earth and Venus.

Central to his whole design, was a golden ratio dividing his sea and sky, and that which coincided with the navel of the goddess, as we saw in Chapter Four, and how would he construct that?

The golden-ratio lines drawn in here onto the picture were drawn by John Martineau. Mathematically-inclined readers will here discern a phi-squared rectangle: putting its width at unity, its

Some 21st-century Golden lines put through *La Nascita di Venere* (Birth of Venus)

height becomes $1+\Phi$. Then the whole picture width is $1+2\Phi$ and dividing by its height gives phi, does it not? (NB, don't worry if you don't get this) Whether or not Botticelli consciously did this, it is assuredly a part of the structure of his picture.

This phi-structure of the picture has of late been endorsed by Gary Meisner in his major work on phi, *The Golden Ratio, the Divine Beauty of Mathematics* – so now it may be official, so to speak. This appears as a new awareness which one will not find discussed prior to the 21st century.

We saw in an earlier chapter how the navel of the goddess is aligned with the horizon, and that gives another phi-ratio in terms of her total height. I suggest that this ratio is basically exact, which cannot be accidental, and indicates that he was working from some sort of canon of 'perfect beauty.' In school maths lessons I used to do, I would pair students to measure each other's height plus their height up to the navel. Averaging all the values together from the class would give quite a good golden-ratio value. I did not have the nerve to put the Botticelli image up on the screen, but in principle would advocate doing this! Phi will always bring a sense of delight into a school maths class, as pi can never hope to do.

Botticelli's image balances the Four Elements, with golden lines dividing Air, Sea and Earth - while the goddess herself is not fiery but awakens the fire of desire in the viewer. Homer called her the Cyprian, and she here lands on Cyprus, her island. The forests of cyprus trees on the island share the etymology, as also does *Cuprum-cuprous*, Cyprus having been the main source of copper in the ancient world - a metal always linked to Venus. The picture glows with gentle copper hues.

If we bisect the picture with a vertical line down the middle, this will pass through the centre of gravity of the goddess, more or less. Aphrodite's weight — if she has any — would pass through that centre line, and that passes just to the right of her foot standing on the conch, the sea-shell. Is she just about to fall over, and would the *Horae* have to quickly catch her? Or, maybe she does not have weight, after all she is the goddess? This is where the Platonic philosophy comes in, with Botticelli having been hovering around the great Italian Neo-platonic school of the Medicis in Florence.

The two *zephyrs* to the left have no weight, because they are the winds, while we see how the *horae* on the other side of the picture stand firmly on solid ground. The roses blown by the *zephyrs* have no roots, rather they are airy and exude a sweet fragrance. On the other left-hand side of the picture, the flowers on the fabric have roots, they are more earthy. We thus feel a balance, between gravity and levity on the two sides of the picture, as the goddess at the centre emerges from eternity into our space-time world:

> *Of august gold-wreathed and beautiful*
> *Aphrodite I shall sing, to whose domain*
> *belong the battlements of all sea-loved*
> *Cyprus where, blown by the moist breath*
> *of Zephyros, she was carried over the waves*
> *of the resounding sea on soft foam.*
> *The gold-filleted Horae happily welcomed her*
> *and clothed her with heavenly raiment.*

> > *Homeric hymn to Venus*, published in Florence in 1488.

There are no flowers on the ground in this picture, nor any oranges on the orange-trees. With her arrival Venus brings flowers and fertility, according to Hesiod. The land is about to blossom, now that She has appeared!

For a thousand years, artists had always painted female figures clothed — or naked just to express the sin, shame and guilt of Adam and Eve after their Fall from primal innocence. Then, suddenly, Botticelli here depicts a naked figure in the very image of divine glory. Here was a tremendously pagan event, with the confidence of a Greek statue. Nothing else in his art was remotely like this - the virgins and angels which he painted were very much covered up!

Yes the picture expressed the yearning of quattocento Florentine women to go blonde - but this image of a naked almost life-size blonde does not appeal to mere sexual desire - for the reason given above, that She does not have weight! This image is about 'desire' in a Latin sense, *de-sidera*, 'from the stars'.

Botticelli was a part of the whole Platonic academy that had formed around Lorenzo de Medici in Florence. It was translating the Platonic and Hermetic texts, which did so much to inspire the Renaissance.

Modern books about phi in art omit mention of Botticelli's *Birth of Venus*, as likewise books on Botticelli omit mention of phi, both missing the essential point: which is, that phi first appears in art, then only a century later finds a mathematic expression.

11

War-God of the Maya

Nobody knows where the people called the Maya of central America came from. They built their temples in the jungle, using neither wheel nor gear. Their mathematics used a zero, with astronomical periods reaching millennia backwards into the past and forging centuries into the future. Only four of their calendar-texts managed to survive the jungle steam plus the zeal of Spanish missionaries. The finest – or the most terrible - of these is called the 'Dresden Codex.' Composed in the eleventh century, it was attuned to the cycles of Venus.

Amidst the huge volcanoes of Aztec and Toltec landscapes, somewhere around the 8th century BC, the 260-day calendar in its combination with the 365-day solar year was set into motion. The Maya calendar has been described as 'the gear- wheels of eternity' - no-one could interrupt its great cycles to reset the wheels. Experts are nowadays piecing together the various surviving Venus codices from this part of the world, to show how these key periods factored in together into their Calendar.

The number 13 keeps turning up in their scheme of things, just like 12 in our calendar system. The 260-day Tzolkin was the engine of destiny generating astrological meanings for each day. Counting to base twenty, it totted up thirteen of them, that being the number of layers of heaven in their cosmology. The Tzolkin is close to the human gestation period, and that is probably the main significance it retains today in Guatemala. Its twenty different day-signs combine with a number 1 to 13, a rolling sequence which has continued without interruption for twenty-five centuries and is still

used by the 'day-keepers' of Guatemala.

A page from the 'Dresden codex'

Our dates have both a day of the month and a weekday, and likewise they had a 20-day 'week,' numbered in a sequence of 1-13.

In the 'Dresden codex' a feathered-serpent Venus-deity rejoices in five different forms or aspects – he had five garbs that change with the eight-year cycle of Venus. These were illustrated in five decorated pages, their version of its eight-year period. Each of the five Venus synodic cycles was thereby perceived in a distinct manner. Each page focussed upon dates when the morning star first appeared, rising before dawn. Those synodic cycles of Venus were further subdivided, with so many days as Morning Star, eight days of disappearance at the inferior conjunction, and so forth.

That codex totted up thirteen of these eight-year cycles, reaching 104 years, or 13 x 13 Venus years, twice 52 years, which was their 'Calendar Round.' This eight-year Venus almanac thus merged into the 104-year 'Great Venus round,' created as the 365-day year and the 584-day synodic period of Venus intermesh with the divinatory cycle of 260 days. Thirteen repetitions of the eight-year period made up this Great Round.

With no leap-year, the Mayan 365-day 'years' followed the 8- year Venus cycle very exactly (to 99.99%), because - as you'll remember - this is always two days short of eight years. In central Mexico '...they counted the days by this star and yielded reverence and offered sacrifices to it,' and knew on what day it would reappear.

Years on the equator are hardly noticed, they are faint, and so the cycles of Venus become the main measure of time-periods, apart from the lunar phases of course.

The Dresden Codex had a table for predicting solar and lunar eclipses, and ephemerides for Mars and Venus. Loads of human sacrifices were linked to these, especially Venus appearing as Morning Star. Twice a year eclipses, solar or lunar, are likely, and an 'eclipse year' keeps in step with these 'eclipse seasons'. Four Tzolkin equal three 'eclipse years'. Every two years and two months Mars glows very brightly in the sky and goes retrograde, the average interval for this, its synodic period, being exactly three Tzolkin, or 780 days. The Tzolkin quantum-interval of the Maya cosmology is multi-purpose, relating to Sun, Moon, Mars and Venus.[20]

The Tzolkin may have been more directly connected to Mars than to Venus, even though their war-schedules were timed by the latter, and Mars had a far lower profile than Venus in their mythology. The Tzolkin period told how long Venus was visible as morning or Evening Star, about 260 days.

A Bloody Hero

The legendary hero Quetzalcoatl the 'once and future king' was represented by a plumed serpent, and his trials and journeys were somehow associated with the cycle of appearances and disappearances of Venus. He made his way to the East coast of Mexico, cremated himself on a funeral pyre, transformed his heart into Venus as Morning Star, then was reborn as risen lord of the Toltecs. Bloody sacrifices honored his progress.

The key event in Maya astronomy was the pre-dawn, first appearance of Venus. This heliacal rising of Venus was the time to go to war: Venus-regulated warfare began by c.250 AD. Their Venus was a male deity whose symbol was a spear-thrower. Murals depict a Venus-enclosure where captives were prepared and sacrificed. The eight-year *Venus War Almanac* was used by warriors of the Tlaloc cult of warfare. It arrived at Tikal as early as the 4th

---

[20] Let's go over the awesome precision here. Three eclipse years (of 346.62 days) equals four Tzolkin to 99.99%. Two Mars synodic cycles equals three Tzolkin to 99.99%. Their 'year' of 365 days links to the Venus-period at the same high-precision, one part in ten thousand; as likewise their round of 104 years equals 13x13 Venus years to 99.97%. These things were all figured out by Geoff stray and J.M.

century AD, where battle captives were sacrificed as players in a ballgame. When the Morning Star appeared in what is now Mexico and Guatemala, houses were closed up to ward off the influence, and captives would then be slain and their blood sprinkled to appease it.

The 52-year 'Calendar Round' of central America signified the 'binding of the years' for the Aztecs, doubling which gives the Huehuetiliztli the 104 great Venus round, used for ritual choreography of war, trade and sacrifice. An Aztec warrior who killed his first captive would get a cloak embroidered with scorpions. Venus was also linked to nourishing rains, fertility and maize, and at the Aztec Atemalqualiztli festival feasting took place every eight years, a time when 'good fortune is sought... and all the gods danced.' Celebrants sang a song relating the maize agricultural cycle to Venus, invoking its birth from the womb of the Earth goddess. Prince Quetzalcoatl brought Maize, and for two days the 'godly dancing' was performed in costumes of the deceased warrior spirits.

The Long Count

These had been four grand epochs or 'Suns,' each ending in cataclysm. The epoch of the 'Fifth Sun' has just ended, but doesn't seem to have ended in cataclysm ...yet.

Almost all the ancient pyramids and stelae in Central America had days inscribed according to this Long Count, composed by the adding of 'baktuns' where a baktun is 144,000 days or nearly four centuries, and thirteen of these span the millennia from August 11, 3114 BC to 21 December, 2012.

That end-date is the consequence of a counting system set in motion more than two millennia ago. Experts are able to interlink Calendar Round inscriptions with this Long Count, although the two measures of time have no inherent connection - they just run along in parallel! The Maya didn't have the word Baktun, that's a modern term, but they did have 'katun' as a bit less than twenty years. There are 260 of these in a 13- baktun great age or 'Sun'. Thus it connects back to their Tzolkin measure of 260 days.

The Maya were masters of calculation, they were the Keepers of Time, nor do they often get the credit they deserve for inventing the zero. Their system of absolute chronology, the so-called long Count, had an initial date corresponding to the mythical day of origin 13 August 3114 BC – not that they were around then, as far as anyone knows. It was used for over a thousand years and ended around the 9th century, when their jungle cities became deserted.

These five 'Suns' measure out the 'Platonic year' – nearly twenty-six thousand years, from the precession of the equinoxes – to within 1% - which is, to say the least, curious.

Recently, the world watched with fascination as the winter solstice of Friday 21st December 2012, whose date was 13.0.0.0.0, turned on the next day into 0.0.0.0.1: a new count, the Epoch of the Sixth Sun. It happened that Earth's seasons then aligned with the galaxy, the Sun's winter solstice position being right on the Galactic equator, a

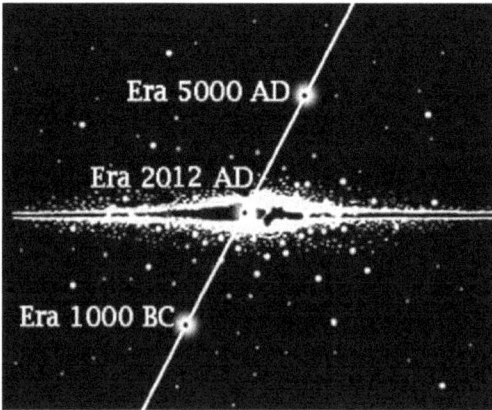

line passing along the centre of the Milky Way, and what was that all about? Surely the Mayans didn't anticipate that, did they? The figure shows how the winter solstice came to align with the Milky Way axis, not far from 2012 …

Figure: the Galactic Alignment of 2012, the first in thirteen thousand years

11

Phi in the Sky

Moving away from Venus now, we focus on the Sun and Moon, and their special secret harmonies. We approach and try to discover in a new manner, the concept of perfection. Let's admit, perfection can seem like a boring and static concept. Centuries ago, the heavens were viewed as perfect and unchanging, whereas here below on Earth things were imperfect, and ever-changing.

Isaac Newton put an end to all that. After him, the cosmic machinery was seen to move by inertia and gravity, and nothing was perfect any more, indeed the word no longer had much meaning. But now, from a different angle, we try to come back, via the 'divine proportion', to some wonderful symmetries. How can a proportion be 'divine'? Botticelli showed us that. In an earlier chapter we mulled over this difficult matter.

Plato is said to have inscribed above the door of his philosophy class in Athens, Let no-one ignorant of geometry enter here. But, he gave no criterion of what one must not be ignorant!

So, let's start off with the diagram here: show that the circle around the pentagram is to the circle inside it, as *the square of the golden ratio*. If that's too hard for you, then this chapter may not be your cup of tea. But, to those who can, the words of Plato in his *Republic* surely apply to them:

> *Then, my noble friend, geometry will draw the soul towards truth, and create the spirit of philosophy.*

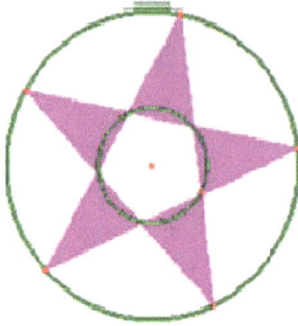

Twenty six centuries ago, he first defined what we can just about recognize as the golden ratio:

It is impossible that the disposition or arrangement of two of anything, so long as there are only two, should be beautiful without a third. There must come between them, in the middle, a bond which brings them into union. The most beautiful of bonds is that which brings perfect unity to itself and the parts linked.. For when of three numbers... the intermediary is to the last as the first is to the last and reciprocally, the last to the intermediary, as the intermediary to the first...

The Timaeus.

Thus, its first definition involved the concepts of beauty and perfection. Such phi-attributes are now as relevant as ever, because of what we are about to discover.

South of the Mediterranean and a century later, Euclid expressed the matter, in more formal terms. After that, nothing much happened until about 1610, when Kepler showed how the Fibonacci series converged towards the Golden Ratio. His colleague Michael Maestlin had written to him a few years earlier, giving phi as a decimal to five or six figures.

Then, towards the end of the 20th century, Robin Heath and his brother Richard applied phi to the solar-lunar measures of time.[21]

---

[21] The Heath brothers discovered these things in 1993: see Robin Heath in *Sun, Moon and Earth*, Wooden Books, 1999, pp.42-3.

Working as a school maths teacher, it always puzzled me that phi was not in the syllabus. I guess that's because of the Newtonian world we have been living in, based on concepts of weight, inertia and gravity. The whole shebang was alleged to have emerged from Chaos, by a chance process. And there was always something inherently suicidal about such a view. But that is all be about to change, and fade behind us into the past. We can awaken to *Kosmos* in the Greek sense which is more focussed upon the concepts of harmony and proportion. Earlier we saw how phi- ratios apply to the planet Venus as our nearest planetary neighbor. We turn to Robin's phi equations. Their precision is compelling.

The Golden Fabric of Time

Phi may not be a 'transcendental' number like pi, but it does go on forever. The Golden Ratio is (as a gnome under a tree explained to me) 'half of one less than the square root of five.' Using it, we here express some basic solar and lunar cycles. They hover around the numbers 18 and 19. Why should that be? The year's length is given by:

$$\textbf{Year} = (18 + \Phi)(18 + 1/\Phi) \text{ days - to } 99.998\%$$

Putting that into words: *eighteen plus phi times eighteen plus the reciprocal of phi yields the solar year,* within ten minutes. Something is holding the length of a year to a golden-perfect length, within ten minutes. Next, what astronomers call the 'eclipse year' is the Sun's revolution against the lunar nodes and it tells us when eclipses will happen. That is quite a bit shorter than the solar year at 346 days, and this too reaches an astonishing precision with its 'Heath formula':

$$\textbf{Eclipse year} = (18 + 1/\Phi)^2 \text{ days - to } 99.997\%$$

Subtracting these two year-formulae tells us that the eclipse year is shorter than the solar year by $18+1/\Phi$ days, or just under 19 days: *the eclipse year is shorter than the solar year by the square root of its magnitude*. Then there is the Islamic year of twelve lunar months, which can also be expressed in these terms:

Islamic Year $= 19(18+1/\Phi)$ days – to 99.8%

but only attains a lower accuracy, of half a day. A thirteen lunar-month 'year' works better:

13 lunar months $= (19+\Phi)(18+1/\Phi)$ days – to 99.99% i.e. within about an hour.

The terms in these formulae are all years. In earlier times when months were lunar, a year contained either twelve or thirteen months. A symmetry is here manifest, in that the thirteen lunar months are greater than the solar year, by the same period as the eclipse year is less.

Take a while to ponder this.

There is also an Indian- Vedic 'Jupiter year,' from Jupiter transiting through one zodiac sign per year, i.e. one-twelfth of its orbit-period. Its called the 'Barhaspatya Samvatsara' and is four days short of the solar year. Again we find a remarkable precision in the formulation given by Richard Heath:

Jupiter year $= 19^2$ days - to 99.99%

These phi-formulae define four different kinds of year to *one part in ten thousand* i.e. within about ten minutes.

We're seeing a connection between length of day, the solar year, and the revolution of the lunar nodes - which ought not to exist according to current cosmology. The period of rotation of the lunar nodes, 18.613 years, slots neatly into this scheme of things.[22] This is

---

[22] The equation here is, 1/eclipse year = 1/year + 1/ node cycle, see Appendix.

the very same 18+1/ Φ but expressed in years not in days: the period by which the eclipse year is shorter than the solar year in days, equals the node rotation period in years. A day-for-a-year concordance is going on here. We may write its equation as:

Lunar node period = (18+ Φ) (18+1/Φ)² days - to 99.97%

- using the 'tropical' value of the lunar node period, the time for the Moon's orbit-plane to revolve once against the zodiac and not its sidereal period (18.599 years). The tropical period has a relation to the Earth, it is more earth-centred.

We earlier found a day-for-year concordance with the Venus-transit period, readers will recall, in relation to the Venus-node (Chapter 6). Something equally weird is happening here in relation to the lunar node period.

We now come to what are called the Saros and Metonic cycles. The 19-year period of Meton has maintained lunar/solar calendars through millennia, while the 18 year, 11.3 day Saros predicts eclipses. As high-precision coincidences these emerge so to speak from the cosmic machinery, poised enigmatically on either side of the 18.6-year node period. They are both 'mere' coincidences and logically, they should not exist![23] Maybe the phi-formulae here involved give a hint as to why they do exist.

The Metonic cycle will track a relatively small number of lunar eclipses, while the Saros tracks long chains of both solar and lunar

---

[23] Improbability of the Saros: "Three monthly cycles that are relevant to eclipses - the synodic month (29.5 days), the apogee-perigee month (27.5 days) and the nodal month (27.2 days) - have to come together, as well as the year, for the Saros to exist. How often do these come together? Concerning the improbability of the Saros eclipse-recurrence, it has been noted: 'though it seemed that a similar eclipse would take place only after an extremely long interval, two prodigious coincidences bring the period to less than 20 years, and make the Saros Cycle a cycle of considerable interest." (Vincent de Callatay, *Atlas of the Moon*, 1964, p.56.) Ptolemy's *Almagest* simply defined it as a period in which these three lunar months coincided. That is the primary definition, while a secondary, unrelated definition puts it equal to nineteen eclipse years, as above.

eclipses. We multiply the solar and eclipse years by 19, to obtain these two periods:

$$\textbf{Saros} = 19(18+1/\,\Phi)^2 \text{ days - to } 99.99\%$$

$$\textbf{Metonic} = 19(18 + \Phi)(18+ 1/\,\Phi) \text{ days - to } 99.998\%$$

Thus we express the Saros as 19 eclipse years, within an hour or two - though this isn't its primary definition*. The ratio between Metonic and Saros periods is normally expressed as that between the 223 and 235 lunar months, whereas it is here viewed in more 'golden' terms as that between $18 + \Phi$ and $18+ 1/\,\Phi$; as likewise the periods of the Saros and the lunar node are related as 19 to $18+ \Phi$.

What do these expressions 'mean'? Surely these formulae indicate a system which is perfect, or which is in some way tending towards perfection. The Golden Ratio has carried that meaning at least since the Renaissance, when great artists imbued it with that significance. The precision of these formulae may point us towards a philosophy of optimism, in a Leibnizian sense. One appreciates that this is not easy. But the cosmos, the *Kosmos*, does not give us any other option. These remarkably exact equations – which should not exist according to everything we've been taught - are concordant with the harmonies of Venus as we have found them. They express some sort of balance of cosmic equilibrium. They indicate that the universe is not a chance affair, but in some way expresses a principle of rationality, which surely has to mean that some ratios are special. They may be the most beautiful equations since Euler discovered that $e^{i\pi} = -1$.

No place like Earth

We have looked at how meaningful patterns appear from an Earth-centred viewpoint, as if telling us that there is something special about that perspective. Well of course there is, in the sense that we live here. The Greek notion of *Kosmos* was based upon the idea of proportion and scale. The Sun and Moon appearing the same size and exactly eclipsing each other has to be our primary, wonder-evoking, experience. A total solar eclipse traces out a thin pencil of darkness that moves across the earth, finely-balanced.

When Isaac Newton developed his astronomy, he used sidereal motion, i.e. rotation or orbit periods with respect to the fixed stars. This gave us the image of a vast universe in which our world is a mere insignificant speck, just whirling... etc., etc. But these expressions are Earth-centred, they don't use star-time. The periods we have here examined are mainly *tropical* and that means, with regard to the Earth-Sun system. It's the music of Gaia. Thus the node-rotation period used above has to be its tropical period, i.e. measured against the zodiac – not against the stars. Its sidereal period won't give us that special concordance. Venus – and also the Moon – give special concordances with respect to Earth.

So these expressions point towards Earth, not the stars, and maybe that is because self-aware consciousness developed here. It is the special nature of the golden ratio that it reflects back upon itself. These equations exclude the notion that our world came about from a chance, random process. They give us instead a more *rational* view, where certain *ratios* give us the reasons for the seasons and why life is here. A hurled world whirled, rotating rightly to remembered rhythm.

We saw how the Fibonacci series journeys ever nearer towards the expression of phi, how plants use it and how the Earth-Venus period-ratios did more exactly fit the Fibonacci values (using 5, 8 and 13) than they did phi. The most precise value achieved on Earth by Mother Nature seems to be in the sunflower, whose spiral-patterns clockwise and anti-clockwise sum to 55 and 89, but do sometimes achieve the Fibonacci 89:144 ratio. That is as exact as any of the formulae we've examined here!

Thus Heaven and Earth compete to express – the Golden Ratio.

Figure: fractal pentagon, made by Roman Chijner, Heidelberg

Data here used: Solar year 365.242 days, eclipse year 346.620 days, synodic month 29.5306 days, 1/12th Jupiter period 360.96 days, synodic period of Jupiter 398.8840 days

12

Man on Mars?

Mars and Venus are opposites. Venus weaves out exquisite proportions in her path of beauty, but rough Mars has none. Venus's surface is unseeable to the human eye, whereas Mars fascinates astronomers by what is seen on its surface. Mars stimulates our will-to-action, - Mars is the *challenge.*

Mars and Venus perform a dance together, as each in turn comes closest to Earth. Each grows brighter in their proximity. We experience this coming-closest very differently: Venus vanishes from view, she is then *invisible* - whereas Mars then glowers most brightly, red in the night sky. One of these cycles takes just over two years and the other, just under. They come nearest to us in a 3:4 rhythm. That is the music of their synodic cycles, and it chimes within an amazing 99.8%.

In winter-time, one-third of Mars' atmosphere condenses into its frozen poles. Huge ravines recall the abundant water which once gushed around, but is now locked underground in deep freeze. There was oxygen in the atmosphere, remembered by all the red iron oxide on its surface. There must have been plenty of carbon, now condensed as dry ice at its poles. Did life once emerge on Mars, when it was warmer and wetter?

Various astronomers have suggested that the seasonally-changing colors of its surface were due to vegetation. Fractal-type images on its surface could be vegetation (these are near the South Pole, where runoff water might be available for them). Mars has a 24-hour day and the tilt of its axis produces changing seasons through its year, no doubt why *Homo sap.* dreams of living out there one day: we'd only weigh half as much.

June 26, 2001

Mars: its big storm of 2001

Red storms rage across Mars. Especially when Mars comes nearest to the Sun, they last for several weeks. Its whole surface becomes covered with pink dust, except for the white, icy poles. Iron dust seethes up covering the highest volcanoes, higher than any on Earth. The NASA image here shows a duststorm developing that enveloped the whole planet: around the September 11th event, which I thought seemed appropriate.

Can we excuse NASA for showing all its Mars images as a dreary orange monochrome, including the sky? When the storms are over, the sky on Mars is blue! Yep it's blue, or maybe a greyish-blue. Search for true-color, Arizona-type landscapes of Mars on the web.

Pictures of alleged blue skies of Mars made newspaper headlines in 2016.

Dawn of a Space Age?

Missions to Mars keep failing, as if jinxed. Of the three dozen Mars-missions so far, no less than two thirds have failed - they blink out as they are getting there, or fail to open up and crash, or miss and go into deep space. Nothing has more knocked the guts out from the dream and the hope of a dawning space age for the human race in this new millennium – than, the continuing, catastrophic failure of missions to Mars. The few probes that do get to Mars always seem to display the same dreary, monochrome stretch of desert. NASA seems reticent about taking pictures of the 'Cydona' region of Mars, which is the part that we, the public, want to hear about.

The glass tunnel on Mars, which convinced Arthur C. Clarke

In 2001 the great sci-fi visionary Arthur C. Clarke became convinced that intelligent life had really existed on Mars: from his inspection of various artifacts, not least an underground 'glass tunnel' appeared – or a wormhole, something indicative of organic life near the Cydona plain, having a 'ribbed cross-structure'. What you see in this image, is several miles long. That, together with other cityscape-type artifacts in the 'Cydona' region were what convinced Sir Arthur C.Clarke.

Today, there is life on Mars. But, don't get too excited, it's probably just dormant bugs, waiting for moisture and warmer weather to start breeding again. Mars does keep producing methane – hundreds of tons a year – and other gases such as ammonia, that are associated with life. "We believe the Viking landers found life on Mars."[24]

Huge amounts of water and dry ice are locked up in its frozen poles, and its daytime temperature hovers around a comfy 15∘C. Its iron-rich soil could probably grow vegetables. A fifty-mile wide crater full of ice was found in 2018. Mars has lost all its atmospheric oxygen, which must have been there once, because of all the oxidized iron on the surface, which makes it 'the red planet'. That's where the dream of 'terraforming' Mars comes in, of electrolyzing the dry ice on the poles, to produce oxygen for the sky and carbon for life.[25] There is plenty of oxygen locked away there … but not much nitrogen, that's the catch.

A 'City on the Edge of Forever'

Mars has two potato-shaped, meteor-battered moons, called Fear and Terror, 'Phobos' and 'Diemos,' named after the section in Homer's Iliad where Mars prepares to emerge onto the battlefield:

And he ordered Phobos and Diemos to harness his horses,
While he himself donned his sparkling armour.

---

[24] *We are not Alone, Why we have already found extraterrestrial life*,' Schultze and Darling, 2009
[25] Heather Cooper and Nigel Henbest, *Mars*, 2001.

Phobos has a very low orbit, and zips round Mars once in seven hours: destined to disintegrate, doomed to destruction, calculated to crash – but, not just yet. A Martian would see these two zoom by in opposite directions every day or two, one rising in the East and the other in the West, even though seen from space they both revolve the same way. Diemos' orbit has zero eccentricity (they put it at 0.0002), so its distance from Mars is fixed, just like the orbit of a modern communications satellite

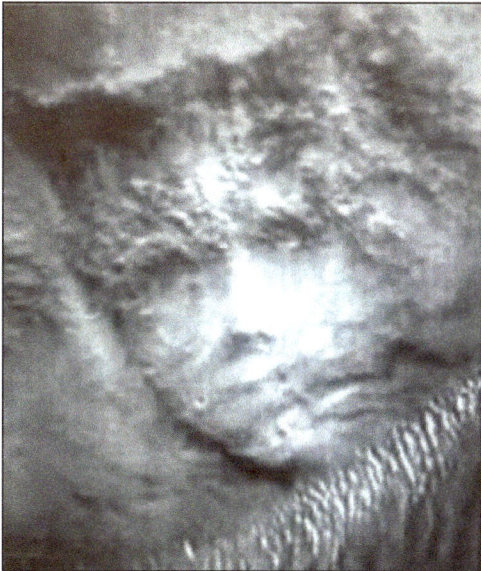

'Egyptian Head' on Mars?

Mars beckons us, to a future that is interplanetary. That is the hope of the human race. NASA needs to be encouraged to go beyond its Brooklins-Institute mentality. This was a report back in the 1960s which warned that contact with an 'other' civilization could undermine that on Earth by somehow demoralizing it. NASA used this to betray the terms of its founding-mission by fudging the data it has really found, of the remnants of some kind of an old civilization on Mars.

To view the pictures, install 'Google Earth' on your computer. Click 'Mars' at the top and it will appear, before your amazed gaze. Here are some co-ordinates[26], input them and it should zoom in to the very spot on Mars where the photo was taken. Put 'Cydonia' into the Search, and checkout 'the Face' (For some intriguing spots to visit, check out the van Flandern 'metaresearch.org' site)

Mars summons us to collaborate in the grand adventure of getting into space. We'd like to see a multicultural team in a spaceship – revolving slowly just as Arthur C. Clarke described, and going to Mars – and maybe lead-clad to protect against radiation. Solar panels would unfold on Mars, for the electricity during one synodic cycle: as the distance Earth-Mars varies so enormously, any visit there would have to take one whole synodic period, or multiples thereof. They'd need cover to avoid getting frazzled by UV as there is no ozone layer in the sky to protect them.

Mars and Venus have no magnetic field, nor does the Moon. This highlights the special uniqueness of Earth having a strong magnetosphere, which is a protective membrane in the space around it. Earth's magnetosphere seems to be generated by the precessional rotation of Earth, caused by our Moon, that stirs up whatever is producing the magnetic field. It's vital for life to exist. Thus we experience the uniqueness of Gaia in space.

This chapter was inspired by Anthony Beckett's talk, 'The Blue Skies of Mars' at the Leeds 2010 Exopolitics conference

---

[26] 'Trees': 284.38°W 82.02°S. 'Tube': 27.98°W 39.12°N. Face: 275.53oW 2.66oN.

Epilogue

The Moon and Sleeping Beauty

Who doth not see the measures of the Moon? Which thirteen times
she danceth every year, And ends her pavan thirteen times as soon
As doth her brother ...   Orchestra, Sir John Davies (1594).

The Moon revolves thirteen times in space each year, whereas
Venus does this much more slowly, revolving only twelve times
within its eight-year cycle. These two traditionally feminine planets
have the most subtle rhythms and harmonies in their motions.
Luna has the same orbit-period as that of its rotation in space, so
that it always faces towards us - whereas Venus, as we saw,
manages an exquisite 13:12 ratio between these two.

The Moon goes round against the stars thirteen times each year,
which is its sidereal period. Also, it moves thirteen times faster than
the Sun, so each day it moves thirteen degrees in the zodiac while
the Sun just moves one degree. So the number thirteen is strongly
lunar, which could be why people are frightened of it.

With the suppression of women's rites we lost touch with the
lunar side of life and thirteen became unlucky. The Sleeping Beauty
fairy-tale describes this. As the cares of the daytime ebbed away,

people would gather around the storyteller, whose long memory would return with the quiet of the night-time, and he would begin:

> Once upon a time, there was a king who had a beautiful daughter, and he wished to invite people to a baptism-feast. There were thirteen wise women (or fairies) in the kingdom, but the king only had twelve golden plates.

So the thirteenth fairy was not invited. Instead, after each of the twelve had given their blessings to the new child, she turned up unannounced and bestowed a curse, concerning how the princess would have to bleed on reaching her fifteenth year.

The king was a sensible fellow who liked the number twelve, because it divided by all sorts of factors, whereas thirteen was an 'irrational' prime number, ugh!

As a reasonable man he wanted his daughter to remain white as a lily, he didn't want her to turn crimson with blood. He named her princess 'Aurora' a solar name, in fact that of the golden goddess of the dawn, just as the plates for his party were made of gold which is the metal that used to belong to the Sun (the Latin word for gold is aurum). No wonder he only had twelve golden plates, expressing the way in which the lunar number thirteen had been ejected from the calendar where it should belong.

It was decreed that there were twelve months only, just as there are twelve hours in the solar day: and these were glued to the solar year, they couldn't move round any more like they used to do, when a thirteenth had to be inserted every few years to keep the year in balance. Thus the king in the story undermined the very fabric of Time, and so it stopped! Everything stopped happening as Time was suspended, until finally a brave prince arrived who was not afraid to bleed, by fighting his way through a thicket of briars (wild roses), to rescue her. Then Time started up again.

But ... did they live happily ever after? Anxious faces gaze at the storyteller, as he pulls from his back-pocket two crumpled-up playing cards, a King and Queen. They are the numbers 12 and 13 in the sequence, he explains, and who says they are in the wrong

Princess Aurora pricks her finger.

order? Personal fulfillment depends upon integration, between the feminine-lunar (thirteen) and the twelve of the Sun's 'golden plates.'

Julius Caesar was stabbed to death on the 'Ides' of March in 44 BC. For a long time, the 'Ides' had signified the period of the Full Moon, at the middle of the month, when sacred ceremonies were held. But, a few years earlier, Julius Caesar had approved of a change in the calendar, so that the months became purely solar, a twelvefold division of the year: so, when he was stabbed to death, the Ides of March no longer signified the Full Moon, but was merely March 15th. Thereby Luna was ejected from the calendar and she could no longer participate in measuring the flow of Time, and it has remained that way ever since.

Quiz Questions:

How many astronauts walked on the Moon?

How many astronauts went there?

How many rockets were built to go there?

Answers:

Twenty-four of them went, three went twice,

so there were 27 astronaut-visits.

Twelve Moon-rockets were built,

but three were cancelled and only nine went.

Each had 3 astronauts,

and six of them touched down on the Moon.

Each time two of the three astronauts landed,

so twelve men walked on the Moon.

Appendices

I: The Connecting Link

There is a handy little equation that will make sense of the periods and cycles here used. To start with the Moon, its two primary cycles are the 27 - day orbit-period, in which it travels once round against the stars, and its 29 - day synodic cycle. The latter one might call the 'blood-rhythm,' because it gives the overall, average length of the female cycle (29-30 days), as well as of the two luminaries meeting in the sky. We can link these two cycles using the year- length:

$$1/27.3 - 1/365 = 1/29.5 \text{ days}$$

If we picture these reciprocal functions as rates of movement, then the orbit speed of the Moon, less that of the 'Sun' (i.e., Earth's period going round the Sun), equals the relative speed of motion between the Sun and Moon. OK?

Next, we answer the question: how often does Venus grow into the bright Evening Star? That interval is its 'synodic' period. We put the two years of Venus and Earth into this same equation, and out pops the answer:

$$1/225 - 1/365 = 1/584 \text{ days}$$

Moving onto Mercury, how long is Mercury's day? To find that, we put its spin period ('axial rotation'), together with its year (its 'orbit period') into this same equation, and we find:

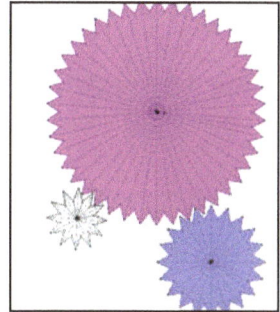

$$\text{Spin period} \quad \text{Year} \qquad \text{Day}$$
$$1/59 \; - \quad 1/88 \; = \quad 1/176 \text{ days}$$

Now maybe you can see why the 'day' of tiny Mercury lasts a huge 176 earth- days. Asimov's robots would have seen the 'terminator,' the day-night boundary, return to the same part of Mercury's rocky ground over this period. The length of Venus' day we likewise derive from its year plus spin-period:

$$\text{Spin period} \quad \text{Year} \qquad \text{Day}$$
$$1/243 \; + \quad 1/225 = \quad 1/117 \text{ days}$$

Voila!

Quiz Question: which has the longer day, Mercury or Venus?

Answer: Mercury. It's day is twice its year – while Venus' day is, roughly, half its year.

II – Avebury's 'Octaeteris'

The first of these equations appears as being cast in stone as it were, at Avebury (in Wiltshire), at the very dawn of British culture and civilization. It's Britain's main lunar temple. By way of meditating upon the two primary lunar cycles, early Brits found it helpful to haul forty-ton stones into perfect circles expressing the two odd-numbers, 27 and 29. Experts reckon the stones were placed at quite regular intervals in these circles, which is why the numbers can be inferred with some confidence, even though many of the stones are now missing. Alas, only two or three remain standing of the 'starry' north wheel of twenty-seven stones, pertaining to the sidereal month, but the impacted sites remain. They were mainly knocked down at the beginning of the eighteenth-century, the so-called 'Enlightenment'. These two wheels are each about a hundred metres across. People visiting Avebury often confuse the remains of these two inner circles with the much larger outlying ring of stones.

It would have been far easier to construct stone circles of 28 or 30, but these were evidently not required. A large precession would have marched up to the giant Avebury ring, via a long snake-like avenue of standing stones, then presumably walked in between these two circles.

This path is spot-on the one-seventh latitude of the Earth within an arc- second. Have a beer at the Red Lion pub there, to mull over these deep issues. Shockingly enough, the forecourt of this pub has the 'starry wheel' of 27 passing directly underneath it, is nothing sacred?

I'll always remember sitting around in a group on the grass at Avebury, with Robin Heath giving his tour. Not to see the 27 and 29 stones as indicating the two primary lunar months, why you'd have to be pretty silly, he explained.

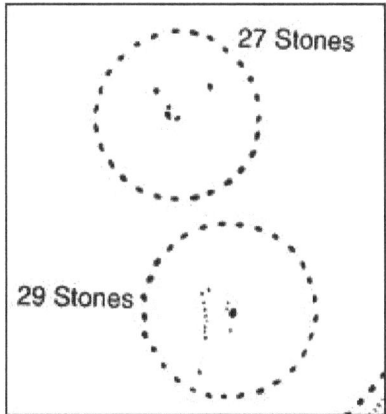

It then took me several years to apprehend, *that no textbooks* alluded to these two numbers in relation to the two now-destroyed

Avebury rings. It's a taboo subject. Even suggesting that the ancient builders could count the number of stones in the rings they made, would nowadays be stepping way out of line. (Even in Robin Heath's *Sun, Moon and Stonehenge* he doesn't mention this rudimentary fact: which links to Stonehenge - or so I used to argue in workshops on the Moon and Stonehenge – because 27 and 29 sum to 56, the number of 'Aubrey Holes' around Stonehenge. That was the beginning of arithmetic, before Sumeria.). Ah, that starry wheel! Stonehenge in contrast has no sidereal connection, it's purely about Sun, Moon and Earth.

There is a chapel on that road between the two lunar-month wheels, ought to be dismantled, because of the large, intact sarsen-stone lying under it, on which its built. That stone needs to be resurrected (as part of the 29-wheel), as likewise should several other intact, buried stones at Avebury. After all this is Britain's primary lunar temple.

Around these giant wheels, there was an even larger ring of stones, and how many of these do you suppose there were? Experts and writers of textbooks are definitely not interested in this question! They will just say, between 97 and 100 stones, and decades roll by with no attempt to check up on this matter – which could easily be done. Likewise they will not even want to commit themselves to the crucial starry-wheel number 27. Robin Heath has argued that the correct number is probably 99 - yes it's the *Octaeteris* again (Chapter 9), the Venus quantum-interval.

If you want to read more about this idea, you'll have to get Robin Heath's book on the subject, *Sun Moon and Stonehenge*, 2002. I doubt whether you will find any other book which mentions it. For the eight-year *Octaeteris* in antiquity, the fullest account in English, or maybe the only decent account seemed to be in Greek Astronomy by Leslie Heath, 1932. It turns out these two Heaths are related, so it's a dynastic thing ... The official map of Avebury counts 98 stones around the perimeter so for a while I was skeptical of Robin's thesis, however it now seems to me likely (I'll spare you the details) that their true number is one more than this. Experts have tended to quote either 100 or 98 as the number of stones around the huge, half-mile perimeter.

Gigantic in design, this monument signifies a primal-beginning of British culture. Who can tell how old it is? Its stones look as if they were carved in some degree, but in weird shapes, redolent maybe of human faces or animal forms. Are they best seen by moonlight? From the dawn of British civilisation comes this stone-age expression of the eight-year Venus period, integrated with the two fundamental lunar months, in a design like two giant gear-wheels.

III - A Century of Venus-Sun Conjunctions

As we saw in Chapter 3, the corners of the Venus-pentagram move very slowly backwards through the zodiac. You can follow this using the longitudes of solar conjunctions given in this Table. The dates of the two Venus-Sun transits are shown in red.

AQUARIUS		SCORPION		VIRGO		PISCES		ARIES	
31.1.50	R10°	13.11.50	21°	3.9.51	R10°	24.6.52	3°	13.4.53	R23°
29.1.54	10°	15.11.54	R 22°	1.9.55	8°	22.6.56	R 0°	14.4.57	24°
28.1.58	R 8°	11.11.58	18°	1.9.59	R 8°	22.6.60	1°	10.4.61	R20°
27.1.62	7°	12.11.62	R19°	29.8.63	5°	19.6.64	R28°GEM	12.4.65	22°
26.1.66	R 5°	8.11.66	16°	29.8.67	R 5°	20.6.68	29°	8.4.69	R18°
24.1.70	4°	10.11.70	R18°	27.8.71	4°	17.6.72	R26°	9.4.73	19°
23.1.74	R 4°	6.11.74	13°	27.8.75	R3°	18.6.76	27°	6.4.77	R16°
22.1.78	1°	7.11.78	R15°	25.8.79	2°	15.6.80	R24°	7.4.81	17°
21.1.82	R 1°	3.11.82	11°	25.8.83	R 1°	15.6.84	24°	3.4.85	R14°
19.1.86	29°CAP	5.11.86	R12°	23.8.87	30°LEO	12.6.88	R22°	4.4.89	15°
18.1.90	R28°	1.11.90	8°	22.8.91	R29°	13.6.92	22°	1.4.93	R12°
16.1.94	27°	2.11.94	R10°	20.8.96	27°	10.6.96	R20°	2.4.97	12°
16.1.98	R26°	30.10.98	7°	20.8.99	R27°	11.6.00	21°	30.3.01	R 9°
14.1.02	24°	31.10.02	R 7°	18.8.03	25°	8.6.04	R18°	30.3..05	10°
13.1.06	R24°	27.10.06	4°	17.8.07	R25°	9.6.08	19°	27.3.09	R 7°
11.1.10	21°	28.10.10	R 6°	16.8.11	23°	5.6.12	R15°	28.3.13	8°
11.1.14	R21°	25.10.14	2°	15.8.15	R23°	6.6.16	17°	25.3.17	R 5°
9.1.18	19°	26.10.18	R 3°	14.8.19	21°	3.6.20	R14°	26.3.21	6°
8.1.22	R19°	22.10.22	29°LIB	13.8.23	R20°	4.6.24	14°	22.3.25	R3°
6.1.26	16°	23.10.26	R1°	11.8.27	19°	1.6.28	R11°	23.3.29	3°
6.1.30	R16°	20.10.30	27°	10.8.31	R18°	2.6.32	12°	20.3.33	R 0°
3.1.34	14°	21.10.34	R28°	9.8.35	17°	29.5.36	R 9°	21.3.37	1°
3.1.38	R13°	17.10.38	24°	8.8.39	R16°	30.5.40	10°	18.3.41	R28°PIS
1.1.42	11°	19.10.42	R26°	7.8.43	14°	27.5.44	R 7°	18.3.45	29°
1.1.46	R11°	15.10.46	22°	6.8.47	R14°	28.5.48	8°	15.3.49	R26°
CAPRICORN		LIBRA		LEO		GEMINI		PISCES	

Ist column: aquarius / Capricorn,   2ⁿᵈ: Scorpio / Libra,   3ʳᵈ Virgo / Leo,
4ᵗʰ Pisces / Gemini,   5ᵗʰ Aries / Pisces

IV - Some 'Islamic' Crescent - Moon / Venus Meetings

The next Table shows those special 'Islamic conjunctions' where the bright Evening Star meets the thin crescent Moon: ideal for that enchanting cocktail-party you were meaning to organize; plus, pre-dawn meetings of the Morning Star with crescent Moon.

118

These dates can be used to checkout the eight-year 'octaeteris' cycle of the ancient Greeks. You'll see how, eight years apart, the

EVENING STAR	MORNING STAR
2005 6/7 Oct, 5 Nov, 4 Dec.	2006 24/25 Feb, 26 March, 24 April.
2007 19 May, 18 June, 17 July.	2007 8/9 Sept, 7th October, 6 Nov.
2009 1 Jan, 30 Jan, 27 Feb.	2009 22 April, 21 May, 19 June
2010 13 Aug, 10/11/12 Sept,	2010 2 Dec, 15 July,
2012 25 Feb, 26 Mar, 24 April.	2011 3 Jan .
2013 8 Oct, 6 Nov, 5 Dec,	2014 29 Jan, 27 March,
2015 21 May, 20 June, 18 July.	2015 9 Sept, 9 Oct, 7 Nov.
2016 3 Dec, 2017: 1 Jan, 31 Jan.	2017 23 April, 22 May, 21 June.
2018 14 Aug, 12 Sept	2018 3/4 Dec, 2019 2 Jan, 1 Feb 219.
2020 27 Feb, 28 March, 26 April	2020 17 July, 15 August.

meetings chime within a day or so. Some dates (shaded in grey) do not look quite so impressive in the sky, e.g. they may not come so close, or may be rather low on the horizon, but they are left in for the sake of showing up this eight-year period.

Index

Ingram Content Group UK Ltd.
Milton Keynes UK
UKHW020429190423
420352UK00010B/52